MW01383388

Springer Series in Reliability Engineering

Series Editor

Professor Hoang Pham
Department of Industrial Engineering
Rutgers
The State University of New Jersey
96 Frelinghuysen Road
Piscataway, NJ 08854-8018
USA

Other titles in this series

The Universal Generating Function in Reliability Analysis and Optimization
Gregory Levitin

Warranty Management and Product Manufacture
D.N.P Murthy and Wallace R. Blischke

Maintenance Theory of Reliability
Toshio Nakagawa

System Software Reliability
Hoang Pham

Reliability and Optimal Maintenance
Hongzhou Wang and Hoang Pham

Applied Reliability and Quality
B.S. Dhillon

Shock and Damage Models in Reliability Theory
Toshio Nakagawa

Terje Aven and Jan Erik Vinnem

Risk Management

With Applications from the Offshore Petroleum Industry

Professor Terje Aven
Professor Jan Erik Vinnem
University of Stavanger
4036 Stavanger
Norway

British Library Cataloguing in Publication Data
Aven, T. (Terje)
Risk management with applications from the offshore petroleum industry. - (Springer series in reliability engineering)
1. Offshore oil industry - Risk management 2. Risk management
I. Title II. Vinnem, Jan Erik
338.2'7282'0684
ISBN-13: 9781846286520

Library of Congress Control Number: 2006940651

Springer Series in Reliability Engineering series ISSN 1614-7839
ISBN 978-1-84628-652-0 e-ISBN 978-1-84628-653-7 Printed on acid-free paper

9 8 7 6 5 4 3 2 1

Springer Science+Business Media
springer.com

Preface

This book is about making decisions in the face of risks and uncertainties. Our ultimate goal is to arrive at decisions that provide desirable outcomes, but the risks and uncertainties oblige us to acknowledge that the best we can do is to obtain *confidence* in being able to obtain desirable outcomes. At the point of the decision-making we do not know the future outcomes of the activities (alternatives) that we are investigating, and the challenge is then, how we should perform the decision-making process. More specifically, some of the main challenges are:

- How we should identify the relevant decision attributes (quantities related to costs, safety, health, *etc.*)
- How we should measure these attributes
- How we should deal with uncertainties in future performance, in general and through different project phases in particular
- How we should balance the project risk management perspective and the corporate portfolio perspective
- How we should take into account the level of manageability in projects
- How we should use expected values in risk management
- How we should understand and deal with risk aversion, the cautionary and precautionary principles as well as the ALARP principle (risk should be reduced to a level that is as low as reasonably practicable)
- How we should formulate and use goals, criteria and requirements to stimulate performance and ensure acceptable safety standards
- How we should understand and use analyses, including risk analyses, to support decision-making
- How we should weight the different attributes, using methods such as cost-benefit analyses, cost-effectiveness analyses and multi-attribute analyses
- How we should involve the stakeholders in the decision-making process.

In this book we address these challenges. A risk management framework is developed providing an adequate context for how to deal with these challenges. The framework comprises

- problem definition (challenges, goals and alternatives)
- stakeholders
- concerns that need to be taken into consideration in the decision-making process
- identification of which risk and decision analyses to execute and execution of these
- managerial review and judgement
- the decision.

Risks and uncertainties are key concepts of this framework. Risk is defined as the combination of the two basic dimensions: (a) possible consequences and (b) associated uncertainties. This definition is in line with that adopted by the UK government (Cabinet Office 2002, p. 7). As there are many facets of these dimensions, the framework implies a broad perspective on risk, reflecting, for example, that there may be different assessments of uncertainties, as well as different views on how these uncertainties should be dealt with.

Compared to much of the earlier discussion on this topic, the book has a higher level of precision with regard to the way uncertainty, probability and expected values are understood and measured. Such a precision level is required to give the necessary sharpness on what risk management can and cannot do.

The framework gives a structure for classification of risk decision problems, and a procedure for execution of the related decision-making processes. The framework provides a check list for what concerns to address when searching for the best decision alternative. Several classification systems are developed, partly based on the risk classification scheme introduced by Renn and Klinke (2002) and modified by Kristensen *et al.* (2005). This scheme is based on characterisations of special features of the consequences, such as ubiquity – which describes the geographical dispersion of potential damages, and persistency – which describes the temporal extension of the potential damages. In addition to this classification of the possible consequences, a system for describing and characterising the associated uncertainties is presented. This system reflects features such as current knowledge and understanding of the underlying phenomena and the systems being studied, the complexity of technology, the level of predictability, the experts' competence, and the vulnerability of the system.

These classifications are based on features of the two dimensions possible consequences and associated uncertainties *i.e.*, the risks. In addition we focus on a third dimension, the level of manageability. Some risks are more manageable than others, meaning that the potential for reducing the risk is larger for some risks compared to others. By proper uncertainty and safety management, we seek to obtain desirable consequences. The expected values and the probabilistic assessments performed in the risk analyses provide predictions for the future, but what the outcomes will be can be influenced. This leads to consideration of, for example, how to run processes aimed at reducing risks (uncertainties) and how to deal with human and organisational factors and obtain a good safety culture.

The framework is presented in Chapter 3, following Chapter 1, which gives an introduction to the book using some real life applications, and Chapter 2 which reviews and discusses fundamental risk management principles and methods. These principles and methods are related to

- basic concepts such as probability, risk, risk analysis and risk management
- economic principles/theories such as portfolio theory, risk aversion, cost-benefit analyses
- basic principles of safety management, such as the ALARP, the cautionary and precautionary principles
- the meaning and use of expected values in risk management
- uncertainty handling (in different project phases)
- risk acceptance and decision-making (risk acceptance criteria, ethical aspects).

The review and discussion are based on recent research related to these topics, clarifying and challenging some of the prevailing paradigms and methods within risk management. Chapter 2 provides new insights and perspectives on basic concepts, theories and methods, and the more practical procedures for implementing them. As an example, we can mention the discussion included on the prevailing approach to the use of risk acceptance criteria and our alternative approach highlighting the generation of alternatives and greater management involvement.

Chapters 4–7 present and discuss several examples of applications, in which the framework of Chapter 3 is used. Chapter 4 focuses on concept optimisation, whereas the operations phase is the main topic of Chapter 5, followed by decommissioning in Chapter 6. An application relating to risk indicators on a national level is discussed in Chapter 7.

In the final chapter, Chapter 8, we discuss the approaches and framework introduced and used in the previous chapters. What are the main challenges and the key success factors? Specifically we address the importance of understanding the basic building blocks of risk analysis and risk management, and challenges related to the implementation of the framework and an ALARP regime.

These applications presume that the reader has a basic knowledge of offshore installations and operations. Although they may lose some details of how the systems being studied work and are operated, readers without such knowledge will also be able to understand and appreciate the main message of these chapters, including the need for risk reducing measures.

Our starting point is the offshore oil and gas industry, but our framework and discussion is to a large extent general and could also be applied in other areas.

This book is written primarily for risk analysts and managers, and others dealing with risk and risk analysis, as well as academics and graduates. To immediately appreciate the book, the reader should be familiar with basic probability theory. One of our goals, however, has been to reduce the dependency on extensive prior knowledge of probability theory. The key probabilistic concepts will be introduced and discussed thoroughly in the book, as well as some basic tools such as cost-benefit analyses. Appendix A reviews basic ideas on risk and risk analysis. This makes the book more self-contained, gives it the required focus with respect to relevant concepts and tools, and opens it up for readers outside the primary target group. The book is based on, and relates to, the research literature in the field of risk and uncertainty. References are kept to a minimum throughout, but readers are referred to the bibliographic notes following each chapter, which give a brief review of the material covered and related references.

The terminology used in this book is in line with the ISO standard on risk management terminology, ISO (2002). Our definition of risk, however, is slightly adjusted compared to the ISO standard, as discussed in Section 2.1. Our focus is the part of risk management addressing HES (Health, Environment and Safety), and in particular major accidents. When we use the term risk management it is tacitly understood that we have in mind these types of risks.

This book is important, in our view, as it provides a guide on how to manage risk and uncertainty in a practical decision-making context and at the same time is precise with respect to concepts and tools in use. Technicalities are reduced to a minimum, ideas and principles are highlighted.

Acknowledgments

Several people have provided helpful comments and suggestions at various stages of the book's development. In particular we would like to mention Kjell Sandve, Henrik Kortner and Vidar Kristensen.

The work is based on a number of research papers and we would like to acknowledge our co-authors; Eirik B. Abrahamsen, Torleif Husebø, Vidar Kristensen, Jens Kørte, Willy Røed, Malene Sandøy, Jorunn Seljelid, Odd J. Tveit, Frank Vollen, Hermann S. Wiencke and Elin S. Witsø.

We would also like to acknowledge the assistance by Evelyn Fulton in order to improve the use of the English language in this book.

Writing this book has been part of the research project HES Petroleum, Decision Tools, managed by the Norwegian Research Council. This project has provided financial support, which is gratefully acknowledged.

The authors wish to thank Anthony Doyle, Oliver Jackson and Simon Rees of Springer-Verlag, London, for their assistance and support in producing the camera-ready version.

Terje Aven and Jan Erik Vinnem
Stavanger, Norway

Contents

1 Introduction 1
1.1 Fundamentals of Risk Management 1
1.2 Challenges 3
1.2.1 Overview of Cases 6
1.3 Summary and Conclusions 17

2 Risk Management Principles and Methods – Review and Discussion 19
2.1 Perspectives on Risk 20
2.2 Economic Principles, Theories and Methods 23
2.2.1 Expected Utility Theory 23
2.2.2 Cost-benefit Analysis and Cost-effectiveness Analysis 26
2.2.3 Portfolio Theory 29
2.2.4 Risk Aversion and Safety Management 30
2.3 The Cautionary and Precautionary Principles 34
2.3.1 Discussion of the Meaning and Use of the Precautionary Principle... 37
2.3.2 Conclusions 40
2.4 The Meaning and Use of Expected Values in Risk Management 42
2.5 Uncertainty Handling (in Different Project Phases) 45
2.6 Risk Acceptance and Decision-making 49
2.6.1 The Present Risk Analysis Regime for the Activities on the Norwegian Continental Shelf 51
2.6.2 A Review of the Common Practice of the ALARP Principle 53
2.6.3 A Structure for a Risk Analysis Regime Without the use of Risk Acceptance Criteria 56
2.6.4 Cases 60
2.6.5 Common Objections to our Approach 64
2.6.6 Conclusions 66
2.7 On the Ethical Justification for the Use of Risk Acceptance Criteria 68
2.7.1 The Influence of the Risk Perspectives Adopted 69
2.7.2 Discussion 72
2.7.3 Conclusions 74

3 A Risk Management Framework for Decision Support under Uncertainty ... 77
3.1 Introduction ... 77
3.2 Basic Building Blocks of the Framework ... 78
3.3 The Framework ... 81
3.3.1 Decision-maker and Stakeholders ... 83
3.3.2 Decision Principles and Strategies ... 83
3.3.3 Decision-making Process ... 83
3.4 Discussion and Conclusions ... 90

4 Applications – Concept Optimisation ... 93
4.1 Historical Background ... 93
4.1.1 Ocean Ranger ... 93
4.1.2 Legislative Situation ... 94
4.2 Typical Current Decision-making ... 95
4.3 Application of the Decision-making Framework ... 96
4.3.1 Framing of the Problem and Alternatives ... 96
4.3.2 Quantitative Results ... 97
4.3.3 Qualitative Evaluations ... 97
4.3.4 Managerial Review and Decision ... 98
4.4 Observations ... 99

5 Applications – Operations Phase ... 101
5.1 Decision-making Context ... 101
5.2 Deficiencies and the Need for an Alternative Process ... 103
5.3 Framing of Decision Problem and Decision Process ... 103
5.3.1 Goals and Criteria ... 103
5.3.2 Problem Definition ... 104
5.4 Generation and Assessment of Alternatives ... 105
5.4.1 Generation of Alternatives ... 105
5.4.2 Assessment of Alternatives ... 105
5.5 Managerial Review and Decision ... 109
5.6 Discussion ... 109
5.7 Observations ... 111

6 Applications – Choice of Disposal Alternative ... 113
6.1 Case Overview ... 113
6.2 Decision-makers and Other Stakeholders ... 114
6.3 Decision Principles and Strategies ... 115
6.4 Framing ... 116
6.4.1 Describe Goals and Objectives ... 116
6.4.2 Problem Definition ... 116
6.5 Generate and Assess Alternatives ... 117
6.5.1 Generate Alternatives ... 117
6.5.2 Selection of Method ... 118
6.5.3 Assess Alternatives ... 118
6.6 Managerial Review and Decision ... 122
6.7 Observations – Decommissioning Phase ... 123

7 Applications – Risk Indicators, National Level 125
7.1 Background and Introduction 125
7.2 Objectives of the Risk Level Project 126
7.3 Overall Approach 127
7.3.1 Major Hazard Risk 127
7.3.2 Other Indicators 128
7.3.3 Leading vs. Lagging Indicators 129
7.4 Event-based Indicators for Major Hazard Risk 129
7.4.1 Indicators for Individual Hazard Categories 129
7.4.2 Basic Risk Analysis Model 130
7.4.3 Challenges in the Trend Analysis 137
7.5 Barrier Indicators for Major Hazard Risk 139
7.5.1 Barrier Elements and Performance Requirements 140
7.5.2 Follow-up of Performance by the Industry 141
7.5.3 Availability Data for Individual Barrier Elements 143
7.5.4 Overall Assessment of Barrier Performance 145
7.6 Observations – Indicators used on National Level 146

8 The Success Factors – Discussion 149
8.1 Understanding the Basic Building Blocks of Risk Analysis and Risk Management 149
8.1.1 Basic Concepts and Theories – Uncertainty 149
8.1.2 Assessments of Alternatives 150
8.1.3 Cost-benefit Analyses and HES 152
8.1.4 Decision Principles and Strategies 153
8.1.5 Research Challenges 155
8.2 Implementation of the Framework 156
8.2.1 Experience with ALARP Demonstration 158
8.2.2 What are the Characteristics of a Good ALARP Demonstration? 158
8.2.3 Needs in order to Improve Applications 159

Appendices
A Foundational Issues of Risk and Risk Analysis 161
A.1 A Wide Spectrum of Risk Indices 161
A.2 Classical, Relative Frequency Perspective 163
A.3 Alternative Bayesian Perspective 164

B Example, ALARP Demonstration 169
B.1 Introduction 169
B.1.1 Purpose 169
B.1.2 Structure of Presentation 170
B.1.3 Basic Assumptions and Limitations 170
B.2 Approach Adopted in ALARP Process 170
B.2.1 Risk Acceptance Principles 170
B.2.2 Illustration of ALARP Process 171
B.2.3 Cost-Benefit Analysis 173

B.3 Risk Results ... 173
B.3.1 Risk to Personnel ... 173
B.3.2 Risk to Assets ... 175
B.4 Identification of Possible Risk Reducing Measures ... 175
B.4.1 Good Practice ... 175
B.4.2 Codes and Standards ... 175
B.4.3 Engineering Judgement ... 176
B.4.4 Stakeholder Consultation ... 176
B.4.5 Tiered Challenge ... 176
B.5 Evaluation of Individual Risk Reducing Measures ... 176
B.5.1 List of RRMs ... 176
B.5.2 Evaluation of RRMs ... 177
B.6 Overall Evaluation of Risk Reduction Measures ... 180
B.7 Final Selection of Risk Reduction Measures ... 183
B.8 Risk Levels after Implementation of Measures ... 184
B.8.1 Measures Accepted for Implementation ... 184
B.8.2 Measures Not Accepted for Implementation ... 185
B.8.3 Residual Risk for Personnel ... 185
B.8.4 Overall Evaluation of Risk ... 186
B.9 Implementation Plan for Measures ... 187

References ... 189

Index ... 197

1

Introduction

In this chapter we first review some fundamental concepts and principles of risk management, as described in the literature and standards, in particular ISO (2005). Then we address and discuss some of the most important challenges in risk management, and point to the need for developing suitable approaches to risk management and appropriate frameworks. Several examples from the offshore oil and gas industry are introduced to clarify some main points and support the conclusions.

1.1 Fundamentals of Risk Management

The purpose of risk management is to ensure that adequate measures are taken to protect people, the environment and assets from harmful consequences of the activities being undertaken, as well as balancing different concerns, in particular HES (Health, Environment and Safety) and costs. Risk management includes measures both to avoid the occurrence of hazards and reduce their potential harms. Traditionally, risk management was based on a prescriptive regime, in which detailed requirements were set to the design and operation of the arrangements. This regime has gradually been replaced by a more goal oriented regime, putting emphasis on what to achieve rather than on the means of doing so.

Risk management is an integral aspect of this goal oriented regime. It is acknowledged that risk cannot be eliminated but must be managed. There is an enormous drive and enthusiasm in various industries and society as a whole nowadays to implement risk management in organisations. There seem to be high expectations that risk management is the proper framework for obtaining high levels of performance.

To support decision-making on design and operation, risk analyses are conducted. The analyses include identification of hazards and threats, cause analyses, consequence analyses and risk description. The results of the analyses are then evaluated. The totality of the analyses and the evaluations are referred to as risk assessments. Risk assessment is followed by risk treatment, which is a process involving the development and implementation of measures to modify risk, including measures designed to avoid, reduce ("optimise"), transfer or retain risk. Risk transfer means sharing with another party the benefit or loss associated with a risk. It is typically effected through insurance. Risk management covers all co-ordinated activities designed to direct and control an organisation with regard to risk, whereas the risk management process is the systematic application of management policies, procedures and practices to the tasks of establishing the context, assessing, treating, monitoring, reviewing and communicating risks, see Figure 1.1.

Risk management involves achieving an appropriate balance between realising opportunities for gains while minimising losses. It is an integral part of good management practice and an essential element of good corporate governance. It is an iterative process consisting of steps that, when undertaken in sequence, enable continuous improvement in decision-making and facilitate continuous improvement in performance.

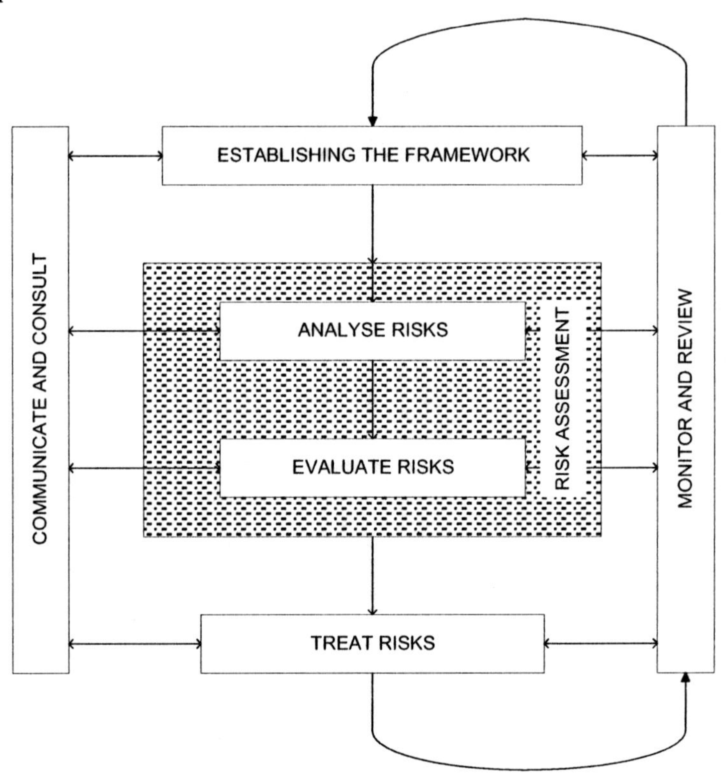

Figure 1.1. The risk management process (based on ISO 2005)

Establishing the context defines the basic frame conditions within which risks must be managed and sets the scope for the rest of the risk management process. The context includes the organisation's external and internal environment and the purpose of the risk management activity. This also includes consideration of the interface between the external and internal environments. The context means definition of suitable decision criteria as well as structures for how to carry out the risk assessment process.

Risk analysis is often used in combination with risk acceptance criteria, as inputs to risk evaluation. Sometimes the term risk tolerability limits is used instead of risk acceptance criteria. The criteria state what is deemed an unacceptable risk level. The need for risk reducing measures is assessed with reference to these criteria. In some industries and countries, it is a requirement in regulations that such criteria should be defined in advance of performing the analyses.

Safety management covers all co-ordinated activities designed to direct and control an organisation with regard to safety. We use the term safety when we focus on risk related to accidents. Hence risk management includes safety management in our terminology. In the literature the terms risk and safety, as well as risk management and safety management are defined in many different ways, and often risk and risk management are used in a narrower sense than here; see Henley and Kumamoto (1983) and Modarres (1993). In safety management emphasis is often placed on aspects related to human and organisational factors, see Mol (2003), Thomen (1991) and OHSAS (2000), in contrast to risk management, which has a tendency to concentrate on the more technical issues.

Similarly we define HES (Health, Environment and Safety) management. Safety management may be seen as a special part of uncertainty management. While uncertainty management considers all uncertainties regarding the project outcome *i.e.*, events with both negative and positive consequences, safety management addresses only the uncertainties that can result in accidents. However, safety management is mainly concerned with low-probability and large-consequence events that are normally not considered in uncertainty management. Hence safety management goes beyond what is typically the scope of uncertainty management (Sandøy *et al.* 2005).

Following our terminology for risk, uncertainty management is a part of risk management, although many aspects normally treated in uncertainty management are not covered by risk management.

1.2 Challenges

Given the above fundamentals of risk management, the next step is to develop principles and methodology that can be used in practical decision-making. This is not straightforward, however. There are a number of challenges and in this book we address some of these:

i) establishing an informative risk picture for the various decision alternatives
ii) using this risk picture in a decision-making context.

Establishing an informative risk picture means identifying appropriate risk indices, and assessing uncertainties. Using the risk picture in a decision-making context means definition and application of risk acceptance criteria, cost-benefit analyses and the ALARP principle (risk should be reduced to a level which is as low as reasonably practicable).

Risk management involves decision-making in situations involving high risks and large uncertainties, and such decision-making is difficult as the consequences (outcomes) of the decisions are hard to predict. A number of tools are available to support decision-making in such situations, such as risk and uncertainty analyses, risk acceptance criteria (tolerability limits), cost-benefit analyses (expected net present value calculations) and cost-effectiveness analyses (addressing, for example, expected costs per statistical saved life). However, these tools do not provide clear answers. They have limitations and are based on a number of assumptions and presumptions, and their uses are based not only on scientific knowledge, but also on value judgements reflecting ethical, strategic and political concerns. Some of the challenges related to these tools are: assessment of uncertainties and assignment of probabilities, determination of appropriate values for quantities such as a statistical life and the discount rate, distinguishing between objective knowledge and subjective judgements, treatment of uncertainties and the way of dealing with intangibles.

Risk analyses, cost-benefit analyses and related types of analyses provide support for decision-making, leaving the decision-makers to apply decision processes outside the direct applications of the analyses. We speak about managerial review and judgement. It is not desirable to develop tools that prescribe or dictate the decision. That would mean too mechanistic an approach to decision-making and would fail to recognise the important role of management performing difficult value judgements involving uncertainty.

Nonetheless, there is a need to provide guidance and a structure for decision-making in situations involving high risks and large uncertainties. The aim must be to obtain a certain level of consistency in decision-making and confidence in obtaining desirable outcomes. Such guidance and structure exist to some degree, and the challenge is to find the right level.

To discuss these issues in more detail we will look at some examples from the offshore oil and gas industry. These address:

1) the early phase of development of the oil and gas activity on the Norwegian Continental Shelf in the 1960s and 1970s
2) year-round drilling and production activities in the Barents Sea
3) evaluation of two field development projects
4) decision-making related to reserve buoyancy for floating installations
5) the operation of installations offshore, national level
6) the operation of installations offshore, installation level
7) decommissioning of offshore installations.

In Example 1 we will discuss the role of risk assessment and management in safeguarding people and the environment. Activities such as diving operations, drilling and transport by helicopter are high risk activities, but the potential benefits (incomes) from the activity are enormous – the interesting question to discuss is then how the various stakeholders (government, oil companies, labour organisations, ...) should face the safety challenges. What should be the basis for determining an acceptable safety level? Although formal risk assessments were at an early development stage at this period of time, or not present at all, we can discuss their potential use if such a tool had been available. And to what extent has the precautionary principle relevance?

Example 2 is similar to the first, but now the field of risk assessments and management has developed into a more mature discipline. Petroleum activities in the Barents Sea have up to now been kept to a minimum, as the area is considered extremely vulnerable from an environmental point of view. The government will accept year-round activities only if the environmental risk has been found acceptable.

Example 3 compares and discusses two different field development projects, subject to reservoir uncertainty and process facility uncertainty. Both projects are associated with risk caused by the uncertain future oil and gas prices.

Example 4 considers decision-making in relation to whether or not to provide reserve buoyancy for floating offshore installations, particularly of the semi-submersible type. Reserve buoyancy is an extra barrier against extreme cases of flooding or severe structural failure.

Example 5 concerns the overall safety level for the total activity on the Norwegian Continental Shelf. What is the present safety (risk) level, and can we see some trends over time? What kind of processes are required to assess and treat the risks for the total activity?

Example 6 discusses how to decide on safety improvements during the operational phase on an offshore installation. A weakness in the protection of personnel in case of major fires has been discovered, and what type of decision-making process should be employed?

Example 7 considers the decision-making process involved in choosing which decommissioning alternative to implement when decommissioning an entire field in the North Sea. Among the alternatives is leaving large structures permanently in place, which has been demonstrated in the past to be a very critical issue.

When deciding whether to start a project, economic evaluations assessing the future economic performance of a project are performed to strengthen the decision basis. The Net Present Value (NPV) is a commonly used economic performance measure in such evaluations. In accordance with the portfolio theory, the NPV analyses focus on expected values and the systematic risk associated with a project *i.e.*, uncertainty in factors that, in addition to affecting the cash flows of the specific project, also affect the cash flows of other projects in the project portfolio. For a well-diversified company or shareholder, the return and the economic risk for the project itself is of course of interest, but more important is the effect this project will have on the return and economic risk for the portfolio as a whole. Unsystematic risk causes some projects to perform worse and others to perform better than their expected values, but, if systematic risk is ignored, the portfolio's result is approximately equal to its expected value. Negative outcomes resulting from unsystematic risk are assumed to be outweighed by positive outcomes in the portfolio.

Uncertainty management techniques are used throughout all phases of a project to minimise delays and to obtain a high performance. And safety management activities seek to identify and implement measures to avoid accidents and reduce accident risk. In contrast to economic project evaluations, uncertainty and safety management analyses mainly consider unsystematic risk, *i.e.*, uncertainty in factors solely impacting the cash flows of the specific project.

Risk and uncertainty are addressed in different analyses and these analyses are performed by different disciplines throughout the phases of a project. However, these analyses adopt different and seemingly conflicting perspectives on risk. While the economic project evaluations focus on expected values and systematic risk, the uncertainty and safety management analyses concentrate on unsystematic risk. But why should we deal with unsystematic risk when it, following the portfolio theory, is not relevant? Why should we identify and control uncertainties in a project when what matters are just the expected values? What is the justification for using substantial resources throughout the various project phases in uncertainty and safety management?

These issues are further discussed in Example 4 and in Sections 2.2–2.5. Of course, analysts and experts from a specific discipline will see beyond the theories developed within their discipline. For example, most economists would find uncertainty management and safety management in a project appropriate even though these activities focus on unsystematic risks. But there is a permanent "conflict of interest" between analysts and decision-makers on a portfolio level and parties on a project level, and to solve this conflict we need to understand the rationale for the various perspectives. To what extent do the portfolio theory and economic cash flow analyses provide the full answers to how to make decisions in projects? Certainly, such theory and analyses are of fundamental importance for obtaining high performance - no one questions this; the issue is rather the constraints and limitations of this thinking, and what the consequences of these constraints and limitations are when it comes to uncertainty and safety management in a project. In Chapter 2 we will clarify and discuss these points. More specifically we will address the following issues:

- To what extent can we ignore unsystematic risks in project management?
- To what extent is the use of expected values relevant and appropriate for steering project performance measures, such as production figures, revenues and number of fatalities?
- What is added by the use of uncertainty and safety management?
- What are the key factors justifying uncertainty management and safety management?
- To what extent are the level of uncertainties and level of manageability important?

1.2.1 Overview of Cases

Example 1: The Early Phase of Development of the Oil and Gas Activity

Let us go back to the early start of the offshore petroleum activities on the Norwegian Continental Shelf in the North Sea in the late 1960s and the beginning of the 1970s. At this time risk analysis was not established as a tool for describing risk, but of course, risk and safety were an issue for the government. An acceptable safety level was required. But what did that mean? There was no explicit description of risk. Well, the answer was proper safety management as performed at that time. And that meant detailed requirements for how to perform and organise activities, such as drilling operations, and the use of supervisory bodies to ensure that the conditions laid down for operations were met. No explicit cost-benefit analysis and utility analyses were performed, but that was not required, as any evaluation of costs and benefits would give a positive conclusion for starting the oil and gas activity. The potential societal benefits would be enormous and would dominate the consequence picture. Accidents could of course occur, but, by introducing a proper safety management system, the government would ensure that safety was controlled in a satisfactory way. The precautionary principle did not apply. Many of the offshore activities were subject to strong uncertainties, such as the consequences of an oil spill and the long-term consequences for divers. Not many people nowadays, however, would say that the government made the wrong decision. Norwegian society would probably have been in a completely different position than it is today, if the government had taken a different approach.

Clearly, responsibility for the many accidents we have experienced falls on the politicians (and we have elected them). No competence in risk assessment would have been required to predict a large number of fatalities (several hundred) and injuries as a result of the offshore activity. Nonetheless, the government initiated the activity. The possible advantages were so great that "they simply had no choice".

There is an on-going discussion on compensation for the divers who were involved in the early phase of development on the shelf. There is no doubt that diving in the North Sea has had its human costs. Many divers have been killed in accidents and many have incurred long-term health problems. The scientific basis for concluding these problems are a direct result of diving can be discussed, but most people would probably consider this not a very fruitful discussion. We cannot disregard those studies and data that indicate a connection between long-term diving performance and failure symptoms of a neurological and cognitive character. This is also the conclusion reached by the commission that recently studied the divers' case. The commission found that the state and the oil companies are legally responsible for the health damage inflicted on the divers. The arguments are clear; those involved were not aware of the possible long-term effects of the diving operations and management of the risk had not been good enough.

An activity was initiated with substantial uncertainties involved concerning the consequences. The pattern is typical. The economic incentives are strong, and uncertainties about the consequences are to some extent suppressed. This latter point is often due to lack of knowledge, but it is also a result of the traditional approach to science; as long as no sufficiently large data set is available to clearly demonstrate a connection between the activity and the damage, people work from the premise that there is no such connection.

The precautionary principle means that we should wait until we have more information and have reduced the uncertainties, but in practice this would be difficult, as the incentives for starting the activities are so strong. But what about the market mechanisms? Should not salaries and other personal benefits compensate for the risks? High risks should mean a high rate of remuneration. Yes, to a large extent this is how our society is organised, but these mechanisms do not work very well in the face of large uncertainties about possible consequences. Insurance policies have a role to play here, but they were not available at that time. The potential for using insurance as an active element in safety management still does not seem to have been fully explored, at least in Norway.

Now, as a thought experiment, let us presume that existing risk management frameworks were available in the 1960s. How would the diving situation then have been approached? Well, undoubtedly, analyses would have been performed covering risk description and risk perception. But how should the risk be described? Following a traditional engineering risk analysis approach, estimates of the real risk would have been computed. Care would have to be shown in dealing with uncertainty. The estimates would obviously be very uncertain as the database would be very limited and the models used to reflect the phenomena under study would not be very accurate. Presenting and communicating risk estimates in such situations has proved difficult, and often the focus remains on the estimates and the uncertainties are more or less ignored. In that way a stronger message from the analysis is obtained, stronger than justified. An attempt to describe the uncertainties may be made, but the problem is that a full uncertainty analysis is extremely difficult to carry out, and in cases like this it would result in very wide intervals for the risks. The message from the analysis is then substantially reduced. The uncertainties become so large that the message is in danger of distortion.

The choice of performance measures is important. Typical candidates would have been:

a) the probability that a diver is killed in an accident in one operation (p_a)
b) the probability that a diver would experience health problems (properly defined) in a certain period of time due to the diving activities (p_b).

In both cases there are uncertainties involved, and most in case b), as the underlying phenomena are not well understood. Let us say that the analysis provides estimates for these two situations that are 1%. For the traditional engineering approach to risk analysis, this number should be discussed in relation to the real risks. But that would result in uncertainty intervals for this number that are extremely wide, for example (50%, 0.05%). The conclusion is that the risks are very uncertain.

The above two probabilities p_a and p_b provide some information, but as the uncertainties are large, other performance measures should also be addressed, for example the proportion of divers incurring health problems. Let X_i be 1 if diver i experiences health problems and 0 otherwise, and let Y be the total number among n that experience health problems. Then we see that Y is the sum of the X_is and the proportion is equal to Y/n.

Now, the risk presentation of the probability in case b), p_b, is equal to the mean probability within the group of divers, that is:

$$p_b = \Sigma_i P(X_i = 1)/n$$

which may also be written as EY/n.

To describe risk, however, it is not sufficient to address mean values, such as EY/n. We need to look at the probability distribution of the proportion Y/n. If there are large uncertainties related to the phenomena involved, this might not be properly reflected in the mean values but in the distribution. In our case, we may for example assign a probability of Y/n greater than 50% to be, let's say, 20%. The point here is not the numbers but the way of thinking. To avoid too detailed analysis we could simply say that there is a large probability, say minimum 10%, of having a significant number of divers experiencing health problems. To obtain a clear message, that might have been sufficiently accurate.

We see that there are many difficult challenges related to assessing uncertainties and expressing risks. There are different perspectives on risks and uncertainties and these have to be understood if we are to be able to obtain the necessary clarification and guidance. We refer to Section 2.1 and Appendix A.

Now let us look at cost-benefit analyses and cost-effectiveness analyses, and suppose that these analyses are to support a decision on getting more information about the possible long-term consequences of diving before further activities are run. Then we need to specify the statistical expected economic loss due to the diving operations, and compare these figures to the expected costs, depending on whether the activities are run as planned or deferred. Clearly, such analyses would conclude that it is not cost-effective to defer activities. The expected cost contribution from possible health problems in the future would be marginal compared to the costs of deferring the activities. No detailed analysis would be required to show this. And, using the economic principle of discounting future costs, the net present value of health problems some 30 years ahead would be negligible.

Similar conclusions would have been reached if cost-benefit analyses and cost-effectiveness analyses had been used to support the decision on starting activities on the Norwegian Continental Shelf.

The analyses thus provide a clear recommendation as to what the best decision is, but the decision-maker naturally has to see beyond the analyses. In the face of large uncertainties, no mechanistic procedure can be adopted based on the use of analyses. Ethical and political aspects need to be considered and to do so we must base our approach on much broader perspectives than formal engineering and economic analyses.

We now leave the pioneering time of the industry and look at a problem relevant today. Norwegian offshore oil and gas activities are now in a different stage, with a safety management system incorporating risk thinking. During the late 1970s and 1980s risk analyses were introduced as a decision-supporting tool and by the 1990s they were an integrated element of the regulations and the safety management systems in the industry. We will address this more in detail in the following example.

Example 2: Year-round Petroleum Activities in the Barents Sea

The Norwegian Government has recently presented a consequence analysis of year-round petroleum activities in the Barents Sea north of Norway. As a part of the analysis a number of sub-analyses have been carried out, including analyses of the consequences related to the environment, employment and fisheries. A special study has been performed on the risk of major hydrocarbon releases to the sea.

The Government and the Petroleum Safety Authority in Norway, which answers to the Government on matters relating to resource management, and safety and working environment for the petroleum activities on the Norwegian Continental Shelf, are positive to continuous operations. The conclusion is that the activity is acceptable from an environmental risk point of view.

But what does this mean? What is the basis? The arguments can be briefly summarised as follows:

The probability of accidental spills in the Barents Sea is no greater than on other parts of the Norwegian Continental Shelf. The physical environment does not present significantly greater technical or operational challenges than players face elsewhere on the Norwegian Continental Shelf.

There is very positive experience during 40 years of petroleum activities on the Norwegian Continental Shelf. Some 1,000 wells have been drilled over almost 40 years – including 61 in the Barents Sea – without any accidental spills which have had environmental consequences. Only one serious blowout has occurred during the operations phase over the same period – the Ekofisk Bravo accident in 1977 – and only one significant spill from oil and gas pipelines. The latter relates to a loading hose left in the open position on Statfjord B in 1992, when 900 cubic metres of oil were discharged. Based on historical records, a probability of 1–10% is calculated for an oil release during the period 2005–2020, depending on future activity level.

Existing technology ensures that the Government's ambition about no releases to sea from drilling operations can be achieved.

In other words, the daily releases to sea will be negligible and the probability for an uncontrolled release is so small that it is judged acceptable.

The government concludes that it is environmental issues that decide whether year-round operations should be accepted or not. Is this really the case? Environmental issues obviously play a key role. But there are no clear answers to what is acceptable from an environmental risk point of view. Is not the point that what is acceptable is related to all aspects of the activity and in particular the benefits that such an activity would generate in terms of income, employment, *etc.*? The judgement is that risk is relatively small and that the positive consequences compensate for this risk. The fact that the conclusion is as it is, is a result of how the consequence aspects are weighted *i.e.*, the issue is more about ethics and politics than technology and science. On a scientific basis, no one can say it is wrong to expose vulnerable areas to possible oil releases by starting year-round operations in the Barents Sea, but this conclusion can be reached through a value judgement.

It is essential to distinguish between facts, analyses and evaluations concerning risk, and value judgements and politics. Facts are related to what has happened *e.g.*, the accidents that have occurred on the Norwegian Continental Shelf and worldwide.

Blowout risk is calculated. The calculations are evaluations based on facts. But the risk numbers are not facts. A risk analysis consultant has expressed the probability of certain accident events occurring, including blowouts. The quantification is based on statistics worldwide. Clearly, evaluations are required to determine the proper population for comparisons. A number of assumptions need to be made to allow us to obtain relevant data and at the same time a sufficient volume of data.

Unfortunately, these two concerns are conflicting. If we restrict attention to the North Sea, the data set is too small. If we include the Gulf of Mexico, we obtain more data, but then the issue is relevance. The consultant makes his best judgement and specifies a probability of 1–10%, based on a number of assumptions. Other assumptions and other consultants would have given other judgements. Perhaps, not so much in this case, but the principle is the important aspect here. The numbers are evaluations and not facts. The consultant indicates that the numbers may be conservative, since they are based on historical data and do not reflect the improvements that have been made and are being made for the activities. Again, this is an evaluation, and not everyone would agree with this conclusion.

The increased focus on cost reduction that we have seen in recent years in the oil and gas industry may have led to a worsening of the safety level. It is also important to note that using statistics may result in low risk numbers, as this does not include contributions from hazards we have not yet experienced.

Now, what do these judgements and evaluations mean? We have to put the numbers and the message from the consequences analyses into a context. We have to interpret the results, and that we do in the light of our value frames and our political goals and ambitions. Then the conclusion may be that the risk is no greater than what we can live with, given the possible implementation of some risk reducing measures. We are facing some possible damage but not loss of irreversible values. In most cases, nature would eventually restore itself. The positive consequences compensate for the accident risk. Alternatively, we may conclude that the risk is large. We would not wish to initiate an activity implying a risk of up to 10% of serious environmental damage in this area.

An important factor influencing our conclusion is our attitude to uncertainty. We do not know what the outcomes of year-round activity in the Barents Sea will be. There is considerable uncertainty as to what will happen. One strategy for dealing with this uncertainty, common among many environmentalists and some political parties, is to completely avoid the activity in such cases. As long as there are considerable uncertainties, let's not take any chances, is the main line of thinking. These concerns may relate to possible environmental damage in the area, but also to an anticipated increase in discharges to the air through increased oil activity. Such a strategy would normally reflect a strong risk aversion attitude and application of the precautionary principle.

Others adopt a more offensive approach to uncertainties and risk. The driving force is what the activities generate of benefit for society and individuals. To make this clear, we cannot build roads in Norway if we do not accept risks. And we cannot build offshore installations if we do not accept that there is a chance of accidents occurring, resulting in fatalities and/or environmental damage. But the added value of these activities is so large that in most cases we would go ahead anyway. If we do not take any risks, there is no life, as all human activities are exposed to possible losses and harm. In a cultural framework, society has strong elements of the entrepreneur prototype.

We refer to Appendix A for further discussions on the issues raised in this example. Section 4.1 considers a more detailed example from the same region, and decision-making relating to optimisation of the production concept for one of the fields in the region.

Example 3: Evaluating Two Different Field Development Projects
Consider an oil company that is evaluating two different field development projects, denoted project I and project II. Both projects concern satellite fields designed to be connected to production installations of already developed fields, they are both in the same economic order of magnitude and have about the same timeframe. Assume that the company will invest in only one of the projects.

There are challenges related to both projects. Project I involves drilling in a reservoir with a narrow pressure margin *i.e.*, during the drilling operations well pressure must be kept within a small pressure interval. Staying within the pressure margin is difficult as the well pressure will not be entirely constant during drilling, but will fluctuate somewhat. The result of the narrow pressure margin can be a blowout and the consequences of a blowout can be large in terms of loss of life, environmental damage and economic consequences. The probability of a blowout is considered small, but there is a limited potential for further reduction.

For Project II, the reservoir conditions are not so difficult, but there are challenges related to the process facility on the existing installations. The well stream from the satellite field can turn out to be substantially different from the well stream from the main reservoir, and this can cause problems. However, if the necessary modifications to the existing process facility are implemented, large problems can probably be avoided.

Both projects are associated with risk caused by uncertain future oil and gas prices, and this uncertainty has a large potential for affecting the profitability of the projects.

The reservoir uncertainty and the process facility uncertainty are considered as unsystematic risk. The difficult reservoir conditions associated with project I will only affect this project. The uncertainty about whether the existing process facility will handle the well stream from the satellite field, will only affect project II. The uncertainty about the future oil and gas price differs to some extent from the reservoir and process facility uncertainties. The oil and gas price uncertainty will affect both projects, and probably also other projects the oil company has interest in. Diversifying against oil and gas price uncertainty is difficult as these prices affect a large part of the economy.

Assume that economic project evaluations are performed for the two projects, in line with the approach traditionally adopted for such evaluations. The result of the analyses is the expected net present values $E[NPV]_{\mathrm{I}}$ and $E[NPV]_{\mathrm{II}}$. The recommendation from the analyses would be to choose the project with the highest $E[NPV]$.

But are the expected *NPV*s sufficient information when deciding which of the two projects to start? Of course not, other factors would be considered, for example the low-probability, but large-consequence event blowout. A blowout will probably have marginal effect on the $E[NPV(r_a)]_{\mathrm{I}}$, but if this event were to occur, the consequences for the project would be large. Can the consequence of a blowout or other events with extreme negative consequences always be outweighed by other projects in the portfolio? And what about the probability and consequences of such events? Low probabilities and large consequences are difficult to assign and the values used are based on a number of assumptions and suppositions. Another analyst or group of analysts may produce other probabilities and consequences and thus a different expected *NPV*. The computation of the expected value of the *NPV* is not an objective process, but depends on the assessors' judgements, and their basis.

And what about the level of uncertainty and the degree to which uncertainty can be affected *i.e.*, the level of manageability? This information is not reflected in the expected *NPV*s. But is the level of uncertainty and manageability important in a decision-making context? If proper uncertainty management offers a large potential for reducing the probability and consequence of process facility problems, is this a reason for choosing project II?

We refer to Section 2.5 for further discussion of this example.

Example 4: Provision of Reserve Buoyancy in Deck Structure

The semi-submersible mobile drilling unit Ocean Ranger capsized on 15.2.82 in Canadian waters. The ballast control room in one of the columns had a window broken by wave impact in a severe storm. The crew had to revert to manual control of ballast valves, but were probably not well trained in this and in fact left the valves in the open position for some time, when it had been assumed that they were in the closed position. Correction of this failure did not occur sufficiently soon to avoid an excessive heel angle. As a result, the rig could not be brought back to a safe state because only one ballast pump room was provided in each pontoon, at one end.

In their regulations from the early 1980s, the Norwegian Maritime Directorate (NMD) stipulated a requirement for reserve buoyancy in the deck, as an extra barrier against capsizing if extensive water filling of several ballast compartments should occur.

Mobile offshore units, including floating production units, have for more than 20 years been designed with reserve buoyancy in the deck structure, in accordance with NMD regulations. Mobile offshore drilling units appear, as a general rule, to be designed according to NMD regulations.

In regard to floating production units of the semi-submersible type, there has for some years been a tendency to question this requirement. Is it really mandatory to install this barrier, or can the regulations be deviated from based upon performance of risk assessments? A few installations have recently been installed without this barrier. We refer to Chapter 4.

Example 5: Status and Trends of the Risk Level Offshore

In 1999, the Norwegian Ministry (Ministry of Labour and Government Administration) responsible for safety asked the Norwegian Petroleum Directorate (now the Petroleum Safety Authority) to develop an approach – a methodology – for assessing the safety level and identifying trends for the offshore oil and gas activities on the Norwegian Continental Shelf. The aim was to characterise the safety (risk) level for the total activities. The purpose was to improve safety by creating a common understanding and appreciation of the safety level and thus provide a basis for decision-making on risk reducing measures. At the time there was much discussion in the petroleum industry as to the actual status of activities in terms of safety and risk. The labour organisations and also others were not very happy with the situation, whereas the oil companies found the safety level very good. It was difficult to see that the conclusions were based on evaluations of the same activity.

The task therefore was partly to establish a common basis, data and methods, for evaluation of the safety level and trends.

This task created a lot of discussion on how to approach the problem. It was not obvious how to solve it. Some people looked for objectivity and methods capable of revealing the truth about safety and risk level. Only if such methods could be developed could consensus be established, it was said. A traditional engineering approach to risk was the thinking, although this was not explicitly articulated. Further discussion, however, revealed that such an approach was not adequate.

The method eventually adopted was to apply an integrated approach, using input from various risk perspectives. Below, we briefly outline and discuss the main features of this approach. We restrict attention to large-scale accidents leading to fatalities.

The starting point for the assessment should be the measurement of some historical accidental events. As far as possible, these data should be objective. It was acknowledged, however, that assessment of the safety level could not be based on hard data only. As safety is more than observations, it was necessary to see behind the data and incorporate additional aspects related to risk perception. A full risk picture cannot be established in an objective way. A broad perspective is required. We need:

- observational data (facts)
- risk analysis descriptions
- perceived risk information
- judgements made by people with special competence
- expert groups
- group of representatives from the various interested parties to build trust and consensus.

Basically, there are three categories of data (which provide different types of information) that can be used:

- losses expressed *e.g.*, by the number of fatalities
- hazardous situations expressed *e.g.*, by the number of major leaks and fires
- events and conditions on a more detailed level, reflecting technical, organisational and operational factors leading to hazards.

But each of the categories shows just one aspect of the total safety picture, and seen in isolation, data from one category could give a rather unbalanced view of the safety level. It was therefore decided that data from all three categories should be incorporated.

A vast number of large-scale accident scenarios could occur in the offshore oil and gas industry, but we have (fortunately) not observed many of these accidents. Using the historical, observed losses, as a basis for the risk assessments could therefore produce rather misleading results. On the other hand, using the events and conditions on a detailed level, as a basis, would also be difficult as such data could be of poor quality. Do the quantities reflect what we would like to address?

Is an increased number of observations a result of the collection regime or the underlying changes in technical, organisational and operational factors?

Hazard measurements were considered to provide the most informative source for assessing the safety level. There were not thought to be any serious measurement problems and the number of observations was considered to be sufficiently large to merit an analysis.

The methods used included interviews and questionnaires designed to elicit information on risk-related behaviour, working environment and conditions, safety management, attitudes and culture, and any underlying factors.

A group of recognised people, with strong competence in the field of risk and safety, was established to evaluate the data observed. These data include the event data and indicators mentioned above as well as other data, reflecting for example the performance of the safety barriers and the emergency preparedness systems. Attention was also given to safety management reviews and results from analysis of people's risk perception.

On the basis of all this input, the group draws conclusions on the safety level, status and trends. In addition, a group of representatives from the various interested parties discusses and reviews important safety issues, supporting documentation and views of the status and trends in general, as well as the conclusions and findings of the expert group. The message from these two groups together provides a representative view on the safety level for the total activity considered. And if consensus can be achieved, this message then is very strong.

Again we are faced with a problem of how to assess uncertainties and describe risk, as well as dealing with the results. A particular approach was adopted in this case, but there are alternatives. What is the rationale for the choice made and what are the challenges related to defining and implementing such an approach? These issues we will discuss further in later chapters, in particular Chapter 7.

Example 6: Safety Improvement of an Installation in the Operational Phase

The case study is related to an existing platform which is part of a so-called "production complex" *i.e.*, with bridge linked installations. The platform in question is a production platform. The scope of the case is the addition of some new production equipment, which will have an impact on risk level. New equipment units means additional potential leak sources, of gas and/or oil leaks which may cause fire and/or explosion, if ignited.

The operator in question had as its sole goal in the present case to satisfy the risk acceptance limits. The FAR value limit (FAR < 10) was rather relaxed and no challenge for the design. The relevant regulations contain requirements for maximum annual impairment frequencies, for certain defined so-called "main safety functions". One of these functions is the need to provide safe escape ways from hazardous areas back to safe areas for a certain period after initiation of an incident or accident. This is called the "escape ways" main safety function. The maximum annual impairment frequency (probability) for main safety functions is $1 \cdot 10^{-4}$ per year. The escape ways may be impaired by several mechanisms *i.e.*, through physical obstructions (blocking) due to severe structural damage, as well as through temporary conditions whereby the escape ways are rendered impassable due to high heat loads and/or dense/poisonous smoke.

The resulting frequency of escape way impairment for the base case design was significantly higher than the acceptance limit, $1 \cdot 10^{-4}$ per year. However, due to the fact that new regulations are not given retrospective applicability in Norway, the current limit for impairment frequency can not be made binding for the installation in question, which was designed several years before the current regulations came into force.

The operator interpreted the values for the impairment frequencies to mean that an acceptable solution would be that no further increase in the escape ways impairment frequencies should result from the addition of new process equipment. A minimum solution to the decision problem was to adopt risk reduction proposals with no really significant effect on safety.

The Norwegian regulations require companies to perform ALARP evaluations, but lay down very few requirements as to how this must be implemented or documented. Many companies will claim that ALARP evaluations have been performed, but there is no documentation of the evaluations that have been carried out. In this case too, the company claimed that ALARP evaluations had been performed. The authorities were not happy with the solution proposed, because it did not address the fundamental issue, but had no legal basis for acting.

The decision to be made is whether or not to install additional fire protection for personnel in order to reduce expected consequences in the rare event of critical fires on the platform. We refer to Chapter 5.

Example 7: Choice of Decommissioning Alternative

This example relates to the choice of decommissioning alternative for offshore installations, using the Frigg facilities as a case.

Offshore structures must normally be removed when the production of field reserves has been completed, according to the OSPAR convention (OSPAR, 1992). Structures exceeding 10,000 tons may on be left *in situ*, depending on a so-called "OSPAR process" based upon acceptance of the solution by all the countries that have signed the treaty. The process further requires recommendation by the relevant national authorities, based upon a comprehensive public hearing of the Environmental Impact Assessment and the Disposal options and plans.

The selection of disposal options for large offshore structures (exceeding 10,000 tons) is a complex process, with many stakeholders. On the one hand we have the operator acting on behalf of the licensees, who have been the owners of the facilities to be disposed of. The Government, on the other hand, will usually cover 70 - 80% of the costs, as a consequence of the tax regimes. Political authorities therefore, explicitly and implicitly, have very strong interests in the decision-making. This is also made very clear through the OSPAR rules for decision-making, where the agreements are between states and not between private companies.

The final decision-makers in this case are the countries that have ratified the OSPAR convention, based upon recommendations from the state with jurisdiction over the structures, UK and Norway in the present case. The Frigg facilities have been a major source of gas production in Europe for about 25 years. The Frigg field straddles the boundary between the Norwegian and the UK Continental Shelves in the North Sea and the operation of Frigg has therefore been in accordance with both UK and Norwegian legislation. Production started in September 1977 and stopped on 26 October 2004.

The public at large may sometimes also become a stakeholder through the involvement of NGOs. This was clearly demonstrated in the Brent Spar case (Greenpeace, 1996), where the public punished the Shell company all over Europe, through different forms of actions against petrol filling stations. We refer to Chapter 6.

1.3 Summary and Conclusions

The above introduction and examples have presented some of the challenges related to risk management, particularly how to assess risk level and how to deal with the risks. The examples and cases will be further discussed in the coming chapters. The established standards such as AN/NZS 4360 (2004), BSI (2000,2001) and ISO (2005), provide a starting point but give no firm guidance on how to deal with the many problems of structuring and implementing risk management. To be able to give such guidance there is a need for a thorough understanding of the rationale of risk management, as discussed in Section 1.2 and summarised in the preface. Chapter 2 provides this rationale, and is the basis for the framework for risk management presented in Chapter 3.

The examples have demonstrated the need for clarification on what risk is, how to express risks and uncertainties, and how to take uncertainties into account in the decision-making processes. It is essential to put the risk assessments and risk treatment in the proper context, to ensure that the different concerns are adequately reflected. This means, for example, that a portfolio perspective is used whenever appropriate. On the other hand, the limitations and constraints of the various theories and methods need to be understood, to avoid a naïve implementation of the tools.

Bibliographic Notes

Examples 1–3 are based on Aven and Kristensen (2005), whereas Example 4 is taken from Sandøy *et al.* (2005). ISO (2005) is the basic reference for Section 2.1.

2

Risk Management Principles and Methods – Review and Discussion

This chapter reviews and discusses fundamental issues in risk management, related to concepts, principles and methods used. We start in Section 2.1 by summarising alternative perspectives on risk, including the prevailing perspectives adopted in engineering, economics and social sciences. A basic distinction is made between the classical approach to risk and probability and the Bayesian approach or paradigm. For readers not familiar with this basic distinction, we refer to Appendix A, which gives a detailed review of these two approaches. Section 2.2 gives an overview of some fundamental economic principles, theories and methods of relevance for safety and risk applications. These include the expected utility theory, cost-benefit analyses, cost-effectiveness analyses and the portfolio theory. This section also discussed the concept of risk aversion in safety management.

Section 2.3 addresses the cautionary and precautionary principles. The "cautionary principle" says that in the face of uncertainty, *caution* should be a ruling principle. The precautionary principle is a special case of the cautionary principle and states that caution should be the ruling principle if there is a lack of *scientific certainty* as to the likely consequences of the action. We discuss how the content and application of the precautionary principle depends on which perspective on risk is adopted.

In Section 2.4 we discuss the use of expected values in risk management. Expected values as a basis for decision-making are supported by the portfolio theory; this is a ruling principle among economists. Many safety experts too view expected values as the key performance measure when making decisions in the face of uncertainties.

Section 2.5 concerns decision-making under uncertainty, throughout various project phases. Different perspectives on risk are used for project management, and seemingly these perspectives are conflicting and not consistent. Portfolio theory justifies the ignoring of unsystematic risk, for example uncertainties related to the occurrence of an accident, whereas in safety management these uncertainties provide an important basis for investments in safety. In addition, uncertainty management is applied to control and reduce risks in the various project phases, focusing on unsystematic risks. In Section 2.5 we discuss this issue. We show that the conflict is based on a lack of precision as to the constraints of the portfolio theory and the ultimate targets for obtaining high performance for the project. Risks reflect uncertainties, and managing these uncertainties is a tool for optimising performance.

In Section 2.6 we review and discuss the use of risk acceptance criteria and the process of making decisions in the face of uncertainties. A basic element in safety management is the use of quantitative criteria and requirements to control risk and safety barrier performance. In this section we challenge this way of thinking. We argue that the prevailing thinking should be replaced by a framework where focus is on the involvement of management in decision-making, through achievement of goals, generation of alternatives and the use of risk analyses, barrier performance analyses and cost-benefit (effectiveness) analyses to compare these alternatives and to the extent possible meet the goals. This means coming closer to the ALARP principle, but is not a direct application of this practice. Challenges related to the practical implementation of such a regime are discussed, in particular the relationship between safety professionals and management, the use of criteria and requirements related to safety impairment loads and barrier performance, the link to industry standards, and the need for involvement by the authorities. The Norwegian offshore oil and gas industry is the starting point for the discussion, but the discussion is to a large extent general. Examples are included to illustrate our way of thinking.

Finally, in Section 2.7, we specifically address the ethical justification of risk acceptance criteria. We conclude that the ethical justification of a regime based on risk acceptance criteria is no stronger than for alternative approaches. Essential for the analysis is the distinction between ethics of the mind and ethics of the consequences, which has several implications that are discussed.

2.1 Perspectives on Risk

A common definition of risk is that risk is the combination of probability and consequences, where the consequences relate to various aspects of HES, for example loss of life and injuries. This definition is in line with that used by ISO (2002). However, it is also common to refer to risk as probability multiplied by consequences (losses) *i.e.*, what is called the expected value in probability calculus. If the focus is the number of fatalities during a certain period of time, X, then the expected value is given by $E[X]$, whereas risk defined as the combination of probability and consequence expresses probabilities for different outcomes of X, for example the probability that X does not exceed 10. Adopting the definition that risk is the combination of probability and consequence, the whole probability distribution of X is required, whereas the expected value refers only to the centre of gravity of this distribution. In the scientific risk discipline there is a broad consensus concluding that risk cannot be restricted to expected values. We need to see beyond the expected values, for example, by expressing the probability of a major accident having a number of fatalities.

Hence risk is seen as the combination of probability and consequence. But what is a probability? There are different interpretations. Here are the two main alternatives:

(a) A probability is interpreted in the classical statistical sense as the relative fraction of times the events occur if the situation analysed were hypothetically "repeated" an infinite number of times. The underlying probability is unknown, and is estimated in the risk analysis.
(b) Probability is a measure of expressing uncertainty as to the possible outcomes (consequences), seen through the eyes of the assessor and based on some background information and knowledge.

Following definition (a) we produce estimates of the underlying true risk. This estimate is uncertain, as there could be large differences between the estimate and the correct risk value. As these correct values are unknown it is difficult to know how accurate the estimates are.

Following interpretation (b), we assign a probability by performing uncertainty assessments, and there is no reference to a correct probability. There are no uncertainties related to the assigned probabilities, as they are expressions of uncertainties.

The implications of the different perspectives are important. If the starting point is (a), there is a risk level that expresses the truth about risk, for example for an offshore installation at a given point in time. This risk level is unknown, true, but in many cases it is difficult to see whether people are talking about the estimates of risk or the real risk.

If the starting point is (b), the experts' position may be weakened, as it is acknowledged that the risk description is a judgement, and others may arrive at a different judgement. Risk estimates also represent judgements, but the mixture of estimates and real risk can often give the experts a stronger position in this case.

Depending on the risk perspective, there may be different approaches to risk analysis and assessments, risk acceptance *etc.* We will discuss this in more detail below; see the following sections.

Seeing risk as the combination of probability and consequence means a quantitative approach to risk. A probability is a number. Of course, a probability may also be interpreted in a qualitative way, using an interpretation such as the level of danger. We may for example refer to the danger of an accident occurring without reference to a specific interpretation of a probability, either (a) or (b). However, as soon as we address the meaning of such a statement and the issue of uncertainty, we must clarify whether we are adopting interpretation (a) or (b). If there is a real risk level, it is relevant to consider and discuss the uncertainties of the risk estimates compared to the real risk. If probability is a measure of the analyst's uncertainty, a risk assignment is a judgement and there is no reference to a correct and objective risk level.

In some cases we have references levels through historical records. These numbers do not however express risk, but they provide a basis for expressing risk. In principle, there is a huge step from historical data to risk, which is a statement concerning the future. In practice, many analysts do not distinguish between the data and the risk derived from the data. This is unfortunate, as the historical data may, to varying degree, be representative for the future, and the amount of data may often be very limited. A mechanical transformation from historical data to risk numbers should be avoided.

There are a number of other perspectives to risk than those mentioned above. Below some of these are summarised (see Pidgeon and Beattie 1998, Okrent and Pidgeon 1998, Aven 2003):

- In psychology there has been a long tradition of work that adopts the perspective to risk, that uncertainty can be represented as an objective probability. Here researchers have sought to identify and describe people's (lay-people's) ability to express level of danger using probabilities and to understand which factors are capable of influencing the probabilities. A main conclusion is that people are poor assessors if the reference is a real objective probability value, and that the probabilities are strongly affected by factors such as dread.

- Economists usually see probability as a way of expressing uncertainty about the outcome, and often in relation to the expected value. Variance is a common measure of risk. Both the interpretations (a) and (b) are applied, but in most cases without making it clear which interpretation is being used. In economic applications a distinction has traditionally been made between risk and uncertainty, based on the availability of information. Under risk the probability distribution of the performance measures can be assigned objectively, whereas under uncertainty these probabilities must be assigned or estimated on a subjective basis (Douglas 1983). This latter definition of risk is seldom used in practice.

- In decision analysis, risk is often defined as "minus expected utility", *i.e.* $-E[u(X)]$, where the utility function u expresses the assessor's preference function for different outcomes x.

- Social scientists often use a broader perspective on risk. Here risk refers to the full range of beliefs and feelings that people have about the nature of hazardous events, their qualitative characteristics and benefits, and most crucially their acceptability. This definition is considered useful if lay conceptions of risk are to be adequately described and investigated. The motivation is the fact that there is a wide range of multidimensional characteristics of hazards, rather than just an abstract expression of uncertainty and loss, which people evaluate in performing perceptions – so that the risks are seen as fundamentally and conceptually distinct. Furthermore, such evaluations may vary with the social or cultural group to which a person belongs, the historical context in which a particular hazard arises, and may also reflect aspects of both the physical and human or organisational factors contributing to hazard, such as trustworthiness of existing or proposed risk management.

- Another perspective, often referred to as cultural relativism, expresses the idea that risk is a social construction and it is therefore meaningless to speak about objective risk.

There exist also perspectives intended to unify some of the perspectives above, see Rosa (1998) and Aven (2003). One such perspective, the predictive Bayesian approach (Aven 2003), is based on interpretation (b), and makes a sharp distinction between historical data and experience, future quantities of interest such as loss of lives, injuries *etc.* (referred to as observables) and predictions and uncertainty assessments of these. The thinking is analogous to cost risk assessments, where the costs, the observables, are estimated or predicted, and the uncertainties of the costs are assessed using probabilistic terms. Risk is then viewed as the combination of possible consequences (outcomes) and associated uncertainties. This definition is in line with the definition adopted by the UK government, see Cabinet Office (2002, p. 7). The uncertainties are expressed or quantified using probabilities. Using such a perspective, with risk seen as the combination of consequences and associated uncertainties (probabilities), a distinction is made between risk as a concept and terms such as risk acceptance, risk perception, risk communication and risk management, in contrast to the broad definition used by some social scientists in which this distinction is not clear.

In this book we adopt a broad perspective, viewing risk as the combination of possible consequences and associated uncertainties, acknowledging that risk cannot be distinguished from the context it is a part of, the aspects that are addressed, those who assess the risk, the methods and tools used, *etc.* Adopting such a perspective risk management needs to reflect this, by

- focusing on different actors' analyses and assessments of risk
- addressing aspects of the uncertainties not reflected by the computed expected values
- acknowledging that what is acceptable risk and the need for risk reduction cannot be determined simply by reference to the results of risk analyses
- acknowledging that risk perception has a role to play in guiding decision-makers; professional risk analysts do not have the exclusive right to describe risk.

Such an approach to risk is in line with the recommended approach by the UK government, see Cabinet Office (2002), and also the trend seen internationally in recent years. An example where this approach has been implemented is the Risk Level Norwegian sector project, see Vinnem *et al.* (2006a, 2006b) and Aven (2003, p.122).

2.2 Economic Principles, Theories and Methods

2.2.1 Expected Utility Theory

The theoretical economic framework for decision-making is the expected utility theory. The theory states that the decision alternative with highest expected utility is the best alternative. The expected utility approach is attractive as it provides recommendations based on a logical basis. If a person is coherent both in his preferences among consequences and in his opinions about uncertainty quantities, it can be proved that the only sensible way for him to proceed is by maximising expected utility. For a person to be coherent when speaking about the assessment of uncertainties of events, the requirement is that he follows the rules of probability. When it comes to consequences, coherence means adherence to a set of axioms including the transitive axiom: If b is preferred to c, which is in turn preferred to d, then b is preferred to d. What we are doing is making an inference according to a principle of logic, namely that implication should be transitive. Given the framework in which such maximisation is conducted, this approach provides a strong tool for guiding decision-makers. Starting from such "rational" conditions, it can be shown that this leads to the use of expected utility as the decision criterion, see Savage (1972), von Neumann and Morgenstern (1944), Lindley (1985) and Bedford and Cooke (2001).

In practice, the expected utility theory of decision-making is used as follows: we assess probabilities and a utility function on the set of outcomes, and then use the expected utility to define the preferences between actions. These are the basic principles of what is referred to as rational decision-making. In this paradigm, utility is as important as probability. It is the ruling paradigm among economists and decision analysts.

Example
We consider a decision problem with two alternatives; A and B. The possible consequences for alternative A and alternative B are $(2, X)$ and $(1, X)$, respectively. The first component of $(\cdot,\cdot)$ represents the benefit and X represents the number of fatalities, which is either 1 or 0. Assume that the probabilities $P(2,0)$, $P(2,1)$, $P(1,0)$ and $P(1,1)$ are

$$\frac{95}{100}, \frac{5}{100}, \frac{99}{100} \text{ and } \frac{1}{100},$$

respectively. The utility is a function of the consequences (i,X), $i = 1,2$, and is denoted $u(i,X)$, and with values in the interval $[0,1]$. Hence we can write the expected utility

$$E[u(i,X)] = u(i,0)\, P(X=0) + u(i,1)\, P(X=1).$$

To be able to compare the alternatives we need to specify the utility function u for the difference outcomes (i,j). The standard procedure is to use a lottery approach as explained in the following.

The best alternative would obviously be (2,0), so let us give this consequence the utility value 1. The worst consequence would be (1,1), so let us give this consequence the utility value 0. It remains to assign utility values to the consequences (2,1) and (1,0). Consider balls in an urn with u being the proportion of balls that are white. Let a ball be drawn at random; if the ball is white, the consequence (2,0) results, otherwise the consequence is (1,1). We refer to this lottery as "(2,0) with a chance of u". How does "(2,0) with a chance of u" compare to achieving the consequences (1,0) with certainty? If $u = 1$ it is clearly better than (1,0), if $u = 0$ it is worse. If u increases, the gamble gets better. Hence there must be a value of u such that you are indifferent between "(2,0) with a chance of u" and a certain (1,0), call this number u_0. Were $u > u_0$ the urn gamble would improve and be better than (1,0); with $u < u_0$ it would be worse. This value u_0 is the utility value of the consequence (1,0). Similarly, we assign a value to (2,1), say u_1. As a numerical example we may think of u_0=90/100 and u_1=1/10, reflecting that we consider a life to have a higher value relative to the gain difference. Now, according to the utility-based approach, a decision maximising the expected utility should be chosen.

For this example the expected utility for alternative A is equal to

$$1 \cdot P(X=0) + u_1 \cdot P(X=1) = 1.0\frac{95}{100} + 0.1\frac{5}{100} = 0.955,$$

whereas for alternative B we have

$$u_0 \cdot P(X=0) + 0 \cdot P(X=1) = 0.9\frac{99}{100} + 0\frac{1}{100} = 0.891.$$

Thus alternative A is preferred to alternative B when the reference is the expected utility.

Alternatively to this approach, we could have specified utility functions $u_1(i)$ and $u_2(j)$ for the two attributes costs and fatalities, respectively, such that

$$u(i,j) = k_1\, u_1(i) + k_2\, u_2(j),$$

where k_1 and k_2 are constants, with a sum equal to 1. We refer to Aven (2003, p. 125).

The expected utility approach is established for an individual decision-maker. No coherent approach exists for making decision by a group. K.J. Arrow proved in 1951 that it is impossible to establish a method for group decision-making which is both rational and democratic, based on four reasonable conditions that he felt would be fulfilled by a procedure for determining a group's preferences between a set of alternatives, as a function of the preferences of the group members, *cf.* Arrow (1951). A considerable body of literature has been spawned from Arrow's result, endeavouring to rescue the hope of creating satisfactory procedures for aggregating views in a group. But Arrow's result stands today as strong as ever. We refer to French and Insua (2000, p. 108) and Watson and Buede (1987, p. 108).

Of course, if the group can reach consensus on judgements, probabilities and utilities, we are back to the single decision-maker situation. Unfortunately life is not so simple in many cases – people have different views and preferences. Reaching a decision then is more about discourse and negotiations than mathematical optimisation.

Decision analyses, which reflect personal preferences, give insights to be used as a basis for further discussion within the group. Formulating the problem as a decision problem and applying formal decision analysis as a vehicle for discussions between the interested parties, provides the participants with a clearer understanding of the issues involved and why different members of the group prefer different actions. Instead of trying to establish consensus on the trade-off weights, the decision implications of different weights could be traced through. Usually, then, a shared view emerges what to do (rather than what the weights ought to be).

We emphasise that we work in a normative setting, saying how people should structure their decisions. We know from research that people are not always rational in the above sense. A decision-maker would in many cases not seek to optimise and maximise his utility, but rather look for a course of action that is satisfactory. This idea, which is often referred to as a bounded rationality is just one out of many ways to characterise how people make decisions in practice.

The expected utility theory is not so much used in practice, as it is difficult to assign utility values for all possible outcomes. The use of lotteries to produce the utilities is the appropriate tool for performing trade-offs, but is hard to carry out in practice, in particular when there are many relevant factors, or attributes, measuring the performance of an alternative.

To make specifications easier, several simplification procedures are presented, see Bedford and Cooke (2001), Varian (1999) and Aven (2003). Nonetheless, the authors of this book still regard the expected utility theory as difficult to use in many situations, in particular for the situations characterised by a potential for large consequences and relatively large uncertainties about what will be the consequences.

It is outside the scope of this book to discuss this in full depth. We refer the reader to Aven (2003). We conclude that even if it were possible to establish practical procedures for specifying utilities for all possible outcomes, decision-makers would be reluctant to reveal these as it would mean reduced flexibility to adapt to new situations and circumstances. In situations with many parties, as in political decision-making, this aspect is of great importance.

Instead it is more common to use a cost-benefit analysis and cost-effectiveness analysis.

2.2.2 Cost-benefit Analysis and Cost-effectiveness Analysis

A traditional cost-benefit analysis was developed for the evaluation of public policy issues. It is an approach designed to measure the benefits and costs of a project, using a common scale. The common scale used is the country's currency. The main principle in transformation of goods into monetary values is to find out the maximum amount society is willing to pay for the project. Market goods are easy to transform to monetary values since the prices of the goods reflect the willingness to pay. The willingness to pay for non-market goods, on the other hand, is more difficult to determine, as discussed below. We use the same example as in the previous section to explain the ideas in more detail.

Example

Two alternatives; A and B are considered. The possible consequences for alternative A and alternative B are $(2, X)$ and $(1, X)$, respectively. The first component of $(\cdot,\cdot)$ represents the benefit and X represents the number of fatalities, which is either 1 or 0. In the cost-benefit analysis we compute the expected monetary values for each alternative, which is equal to $i - E[c(X)]$, where i is the benefit, which is 1 or 2 depending on the alternative, and $c(X)$ is the cost of X fatalities. To determine

$c(X)$, the common approach is to specify the value of a statistical life *i.e.*, the amount society is willing to pay to reduce the expected life by one. Suppose that a value of 2 million USD is used. Then we compute the following expected values for the two alternatives

A: $2 - 2 \cdot (5/100) = 1.90$

B: $1 - 2 \cdot (1/100) = 0.98$.

Hence alternative A is preferable. To change this conclusion a statistical life needs to be higher than 25 million USD.

We leave the example and return to the general theory. To determine a value of a statistical life, different methods can be used. Basically there are two categories of methods, the revealed approach and the questionnaire approach. In the former category, values are derived from actual choices made. A number of studies have been conducted to measure such implicit values of a statistical life. The costs differ dramatically, from net savings to costs of nearly 100 billion USD. Common reference values are in the area 1–20 million USD. The latter category, the questionnaire approach, is used to investigate individual tendency towards risk taking and willingness to pay under different hypothetical situations, see Nas (1996) and Jones-Lee (1994).

Although cost-benefit analysis was originally developed for the evaluation of public policy issues, it is also used in other contexts, in particular for evaluating projects in the private sector. The same principles apply, but using values reflecting the decision-maker's benefits and costs, and the decision-maker's willingness to pay. In the following, when using the term cost-benefit analysis, we also allow for this type of application.

In practice we need to take into account time and the discounting of cash flow, but the above calculations show the main principles of this way of balancing cost and benefit. When taking into account time, we compute the expected net present value, the $E[NPV]$. To measure the NPV of a project, the relevant project cash flows (the movement of money into and out of your business) are specified, and the time value of money is taken into account by discounting future cash flows by the appropriate rate of return. The formula used to calculate NPV is:

$$NPV = \sum_{t=0}^{T} \frac{X_t}{(1+r_t)^t},$$

where X_t is equal to the cash flow at year t, T is the time period considered (in years) and r is the required rate of return, or the discount rate, at year t. The terms capital cost and alternative cost are also used for r. As these terms imply, r represents the investor's cost related to not employing the capital in alternative investments. When considering projects where the cash flows are known in advance, the rate of return associated with other risk-free investments, such as bank deposits, makes the basis for the discount rate to be used in the NPV calculations. When the cash flows are uncertain, which is usually the case, they are normally represented by their expected values $E[X_t]$ and the rate of return is increased on the basis of the Capital Asset Pricing Model (CAPM) in order to outweigh the possibilities for unfavourable outcomes, see Copeland and Weston (1998). Alternatively, the rate r is unchanged and the value X_t is replaced by its safety equivalent c *i.e.*, a value that is known with certainty. At the assessment point, the assessor is indifferent with respect to receiving X_t or c.

In a traditional cost-benefit analysis all attributes should be included in the analysis so that the conclusions of the analysis can give clear answers on which alternative should be chosen. The analysis is based on an idea that there exist "correct" input values for all attributes, for example a statistical life. The correctness refers to the amount society (the decision-maker) is willing to pay for the value. Use of cost-benefit analysis ostensibly leads to more "efficient" allocation of the resources by better identifying which potential actions are worth undertaking and in what fashion. By adopting the cost benefit method the total welfare is optimised. This is the rationale for the approach.

The method is not simple to carry out, as it requires the transformation of non-economic consequences, such as expected loss of lives and damage to the environment, to monetary values. To avoid the problem of transformation of all consequences to one unit, it is common in many situations to perform a *cost-effectiveness analysis*. In such analyses, indices such as the expected cost per expected saved lives are computed. For the above example, this index is given by the expected cost per expected saved life, by going from alternative A to B; *i.e.*

$$(2-1)/[(5/100)-(1/100)] = 25,$$

as the cost difference is 2−1 and the reduction in expected number of fatalities is equal to 5/100−1/100. Hence the cost is equal to 25 million USD per saved expected life. If we find this number too high to be justified, the analysis would rank alternative A before alternative B.

A More Pragmatic View on a Traditional Cost-benefit Analysis

A more pragmatic view on cost-benefit analysis differs from a traditional cost-benefit analysis in two areas. The first difference is that some non-market goods can be excluded from the analysis. This may be done for some attributes for which it is difficult to assess a proper value, such as environmental issues.

The second difference is that there is no search for correct, objective values. Searching for these values is meaningless, as such numbers do not exist. As an example consider the value of a statistical life. This value represents an attitude to risk and uncertainty, and this attitude may vary and depend on the context. Instead the sensitivity of the conclusions should be demonstrated by presenting the results of the analysis as a function of the assumptions made.

A result of these considerations is that a cost-benefit analysis provides decision support and not hard recommendations. The analysis must be reviewed and evaluated, as we cannot replace difficult ethical and political deliberations with a mathematical one-dimensional formula, integrating complex value judgements.

Multi-Attribute Analysis
A multi-attribute analysis is a decision support tool analysing the consequences of the various measures separately for the various attributes. Thus there is no attempt made to transform all the different attributes in a comparable unit. In general the decision-maker has to weight non-market goods such as safety and environmental issues with an expected net present value, *E*[*NPV*], calculated for the other attributes (market goods) in the project. An alternative way to weight the different attributes is to use different ratios, based on a cost-effectiveness analysis.

Cost-benefit analyses used in the more pragmatic way may be a part of a multi-attribute analysis.

2.2.3 Portfolio Theory

The portfolio theory introduces the concepts of systematic and unsystematic risks, and justifies the ignoring of unsystematic risk – the only relevant risk is the systematic risk associated with a project. This is explained in more detail in the following.

Generally, a portfolio consists of *N* different projects. Assume that each of the *N* projects has a 1/*N* weight in the portfolio and let us use the notation $E_i = E(r_i)$ for the expected value of the return r_i and for the variance, $VAR_i = VAR(r_i), i = 1,2,..., N$. Then for the portfolio *p*, the expected return and variance are given byequal:

$$E_p = \sum_{i=1}^{N} \frac{1}{N} E_i = \frac{1}{N} \sum_{i=1}^{N} E_i$$

and

$$VAR_p = \sum_{i=1}^{N} \left(\frac{1}{N}\right)^2 VAR_i + \sum_{i=1}^{N} \sum_{j \neq i, j=1}^{N} \left(\frac{1}{N}\right)^2 COV_{i,j}$$

$$= \frac{1}{N} \overline{VAR} + \left(1 - \frac{1}{N}\right) \overline{COV} \tag{2.1}$$

where

$$COV_{ij} = E\{(r_i - E_i)\ (r_j - e_j)\}$$

and

$$\overline{VAR} = \frac{1}{N} \sum_{i=1}^{N} VAR_i \text{ and } \overline{COV} = \frac{1}{N^2 - N} \sum_{i=1}^{N} \sum_{j \neq i, j=1}^{N} COV_{i,j}$$

We refer to the terms of formula Equation 2.1 as the unsystematic risk and the systematic risk, respectively. The portfolio's actual value is equal to its calculated statistical expected value plus risk, the unsystematic risk and the systematic risk. The systematic risk relates to general market movements, for example caused by political events, and the unsystematic risk relates to specific project uncertainties, for example accident risks. When the number of projects is large, we see from Equation 2.1 that the variance for the portfolio is approximately equal to the average covariance, and each individual variance is not relevant. Thus the unsystematic economic risk is negligible when N is sufficient large. By *diversification* of the risks into many projects, the unsystematic risks are removed. The company's total cash flow (all projects are included) is approximately equal to the expected cash flow to all projects, if the systematic risk is ignored. The relation between the portfolio's actual value (Y') and its calculated statistical expected value (EY') is given by

$$Y' = EY' + \text{systematic risk} \qquad \text{or} \qquad \text{systematic risk} = Y' - EY'.$$

The difference between the portfolio's actual value and its calculated statistical expected value is, from portfolio theory, just dependent on the systematic risk.

If a company, or the owners of a company, are assumed to have invested in a number of different projects, they are well-diversified owners of a portfolio of projects. In accordance with the portfolio theory, when deciding on a project or selecting among projects, only the systematic risk should be considered. This means that the risk-adjusted rate r to be used to determine the expected net present value $E[NPV]$ is only supposed to be adjusted for systematic risk associated with the project *i.e.*, uncertainty in factors affecting all projects in the portfolio. Unsystematic risk, such as accident risk, is not to be taken into account when determining the appropriate risk-adjusted discount rate.

We will discuss the implications of the portfolio theory in the following Sections 2.4 and 2.5.

2.2.4 Risk Aversion and Safety Management

The concept of risk aversion is widely used to describe an attitude to risk and uncertainty. Intuitively we know what this concept means – we dislike negative consequences or outcomes so badly that we give these outcomes more weight than a statistical mean value approach would give. An example makes this clear.

A house is considered to have a value of 1 million. The probability of a fire resulting in a total loss is $1 \cdot 10^{-5}$ for a period of one year. This gives an expected value of 10. The insurance premium is 100. Thus if the house owner is willing to pay the insurance premium, we have a situation where the house owner is risk averse. The house owner pays more than the expected value and is consequently risk averse. The cost of 100 for one year is considered an acceptable price to pay to obtain full compensation in the case of a total loss.

For the insurance company the extent of risk aversion is obviously of interest as it provides a basis for specifying the premium. But is risk aversion of any interest for the house owner? No, it is not. The house owner's attitude to risk and uncertainty is not at all based on a reference to a statistical expected value. For the house owner, the key aspects to consider are the possible values at stake, and the associated uncertainties. Mean values of large populations or centre of gravity of uncertainty distributions do not provide the house owner with much information for determining what he or she is willing to pay for an insurance policy covering the potential full loss of his or her house. Yet risk aversion is often used as a way of explaining why people insure their houses.

The same type of observation is made for many other situations involving safety – safety people often refer to risk aversion as an argument for some specific decisions under uncertainty. If the concept risk aversion simply means disliking risk, this way of speaking is of course correct. However, the term has an alternative definition in a decision analysis context as explained above, and following this definition we should not use risk aversion as an argument for a certain type of behaviour. The concept of risk aversion is a concept that describes rather than determines attitudes. The main reason for investing in safety is not risk aversion (when referring to the concept of risk aversion in the following we will always think of the above decision analysis definition), but the fact that we wish to protect some values in the face of uncertainties – the thinking is cautionary. We invest in safety to reduce uncertainty and provide assurance if a hazardous situation should occur. We may dislike the possible occurrence of some extreme outcomes so much that we are willing to use substantial resources to avoid these outcomes.

Using the term risk aversion means that we have to relate our risk attitude to the expected value. In a safety context our focus is attitudes to uncertainties and risks, but we would not always see these in relationship to the expected value. This makes reference to the risk aversion concept difficult. Note that the expected value is not a unique objective quantity. Different assessors would normally produce different expected values or estimates.

Furthermore, to apply the concept of risk version we need a common scale. This is often difficult in a safety context. The potential consequences are not easily transformed to such a scale. For example, if an activity can result in fatalities or damage to the environment, how should these consequences be expressed on a common scale? There are no unique economic numbers expressing the values of human beings and the environment.

Risk aversion is thoroughly discussed in the literature, mainly by economists and decision analysts. In safety literature, risk aversion is often referred to as an attitude to risks and uncertainties (see Aven, 2003 and Vinnem, 1999), but there seems to be a gap between the theory developed in the economic and decision analysis literature and its practical use in safety contexts. Safety people often lack a proper understanding of what risk aversion really means.

Before we explore the topic in more detail we will introduce and discuss an example.

Example: Year-round Petroleum Activities in the Barents Sea

We return to the Barents Sea example introduced in Section 1.2.

In December 2003 the Norwegian government considered whether year-round operation should be allowed for the ecologically sensitive Lofoten and Barents Sea areas, of large importance both to fisheries, and the oil and gas industry. The result of the process was that operations were allowed in the main part of the Barents Sea. In the Lofoten area petroleum activities were not allowed due to the area's importance as a spawning ground for valuable species of fish and hence its importance for the fisheries. Some political parties and other groupings were against any activity in the area. Was their resistance caused by risk aversion? And what about the decision not to allow activity in the Lofoten area: was this decision due to risk aversion? To further discuss the influence of risk aversion, an understanding of what risk aversion means in this context is needed.

To support the government's decision an assessment of the consequences of year-round oil and gas activities was performed. Below is an example illustrating how such an assessment can be performed. Note that the example is far less extensive and the values used may deviate strongly from the actual assessment used to support the government's decision in 2003.

To assess the consequences of year-round operation in the Lofoten and Barents Sea area, quantities or performance measures that summarise the successfulness of the activity should be identified. Let us for the sake of simplicity say that only two performance measures were considered necessary to summarise year-round operation in the specific area:

Z - seabird life in year s
Y - the net present value of the investments in the area.

Let us consider seabird life Z in more detail.

Assume that in order to quantify the seabird life, a scale from 0 to 1 was constructed. A Z value equal to 1 corresponds to the current level and 0 corresponds to no seabird life, $Z \in [0,1]$. The uncertainty about the value of Z was assessed by use of a probability density $f(z)$. A reduction of Z is possible if a large oil leak occurs. Let us say the experts expected the quantity Z to be 0.4 if a large oil leak occurred and 1 if not. Further assume that the probability of a large oil leak was assigned to be $1 \cdot 10^{-3}$ in the period before year s. The expected value EZ in year s is then:

$$P(\text{ large oil leak before s }) \cdot E(\, Z \mid \text{large oil leak }) +$$
$$P(\text{ no large oil leak before s }) \cdot E(\, Z \mid \text{no large oil leak })$$
$$= 1 \cdot 10^{-3} \cdot 0.4 + 0.999 \cdot 1 = 0.9994.$$

Seabird life in year s will most likely be at the same level as today, however, there is a probability of $1 \cdot 10^{-3}$ for a reduction of Z to 0.4.

Assume that an assessment of Y, the net present value of the investments in the area, was also performed, and that the assessments of Z and Y were used to support the decision to allow year-round operations. What does risk aversion mean in this context?

Let us say that one person is risk averse with respect to the seabird life Z. The definition of risk aversion is that the safety equivalent $C(Z)$ is smaller than the expected value EZ. To illustrate, if the safety equivalent $C(Z)$ is 0.95, this means that the situation of allowing year-round operations in the Lofoten and Barents Sea area is compared with having a reduction of seabird life in year s to 0.95 with certainty. To allow an activity that can result in a reduction to 0.4 is seen as equally negative as having a reduction to 0.95 with certainty.

Discussion and Conclusion

As mentioned above, risk aversion means that we dislike negative consequences or outcomes so badly that we give these outcomes more weight than an expected (statistical mean) value approach would give. But whose expected value is the starting point for the risk aversion? In the Barents Sea example above the expected value EZ was 0.9994. But EZ is not a true, objective value. In mathematical terms the expected value is written $E[Z|K]$, where K is the background information. Different experts have different background information, and the results may differ substantially. In the definition of risk aversion we compare the safety equivalent with an expected value. The basis for the expected value is often statistical material, or analyses and evaluations from experts. But a true, objective expected value does not exist. For example, another expert group can look at the same situation from a different point of view and determine an expected value for the Barents Sea example equal to 0.90 instead of 0.9994. If a person then specifies a safety equivalent of 0.95, he will be a risk seeker if 0.90 is the reference value, but risk averse if 0.9994 is used. We see that the conclusion that the person is risk averse becomes rather arbitrary.

But even if we could agree upon a specific expected value EZ, the reference to this value cannot be used as an argument for our preferences. In the Barents Sea example, most people would have a safety equivalent less than the expected value, and thus be risk averse. Knowing the person's safety equivalent is more informative. The safety equivalent provides information about the person's attitude to uncertainty and weight of the possible negative outcomes.

However, neither risk aversion nor the safety equivalent gives clear recommendations on whether or not to allow year-round operation in the Barents Sea area. Even if someone is extremely risk averse, he or she may still be in favour of year-round petroleum activities in the Barents Sea. The point is that the economic advantages compensate for the environmental risk. Of course, if year-round operation in the Barents Sea area means a reduction of seabird life to a very low value, say 0.50, with certainty, there must be very strong economic advantages to compensate for the environmental risk. But if the economic benefits are sufficiently large, the person will probably be in favour of allowing the activity.

Risk aversion is a way of characterising behaviour under uncertainty, but cannot be used to justify decision preferences. There are other factors to take into account than those reflected by the risk aversion concept.

To conclude, risk aversion is not the correct term to explain a specific stance in decision problems involving safety, for example that you are not in favour of allowing year-round activities in the Barents Sea area. The explanation is cautionary or because you find that the benefits of the activities do not compensate for the environmental risks. There is little value to be added by discussing whether your safety equivalent is less or greater than the expected value, as the expected value is not an objective quantity. The concept of risk aversion is a theoretical concept characterising preference behaviours, but cannot be used to predict preference behaviour. This is a well known fact among economists and decision analysts, but seems to be overlooked by many safety people.

2.3 The Cautionary and Precautionary Principles

The cautionary principle is a basic principle in safety management, expressing the idea that, in the face of uncertainty, *caution* should be a ruling principle. This principle is being implemented in all industries through safety regulations and requirements. For example in the Norwegian petroleum industry it is a regulatory requirement that the living quarters on an installation should be protected by fireproof panels of a certain quality, for walls facing process and drilling areas. This is a standard adopted to obtain a minimum safety level. It is based on established practice of many years of operation of process plants. A fire may occur, it represents a hazard for the personnel, and in the case of such an event, the personnel in the living quarters should be protected. The assigned probability for the living quarters on a specific installation being exposed to fire may be judged as low, but we know that fires occur from time to time in such plants. It does not matter whether we calculate a fire probability of x or y, as long as we consider the risks to be significant; and this type of risk has been judged to be significant by the authorities. The justification is experience from similar plants and sound judgements. A fire may occur, since it is not an unlikely event, and we should then be prepared. We need no references to cost-benefit analysis. The requirement is based on cautionary thinking.

Risk analyses, cost-benefit analyses and similar types of analyses are tools providing insights into risks and the trade-offs involved. But they are just tools - with strong limitations. Their results are conditioned on a number of assumptions and suppositions. The analyses do not express objective results. Being cautious also means reflecting this fact. We should not put more emphasis on the predictions and assessments of the analyses than can be justified by the methods used; refer to the discussion in Abrahamsen *et al.* (2004).

In the face of uncertainties related to the possible occurrences of hazardous situations and accidents, we are cautious and adopt principles of safety management, such as

- robust design solutions, such that deviations from normal conditions are not leading to hazardous situations and accidents,
- design for flexibility, meaning that it is possible to utilise a new situation and adapt to changes in the frame conditions,
- implementation of safety barriers, to reduce the negative consequences of hazardous situations if they should occur, for example a fire,
- improvement of the performance of barriers by using redundancy, maintenance/testing, *etc*.
- quality control/ quality assurance,

- the precautionary principle, saying that in the case of lack of scientific certainty on the possible consequences of an activity, we should not carry out the activity,
- the ALARP-principle, saying that risk should be reduced to a level which is as low as reasonably practicable.

The level of caution adopted will of course have to be balanced against other concerns such as costs. However, all industries would introduce some minimum requirements to protect people and the environment, and these requirements can be considered justified by reference to the cautionary principle.

In this section we will draw special attention to the precautionary principle, whereas the ALARP principle will be discussed in Section 2.6.

There are many definitions of the precautionary principle; see Lofstedt (2003) and Sandin (1999). The most commonly used definition is probably the 1992 Rio Declaration:

> In order to protect the environment, the precautionary approach shall be widely applied by States according to their capabilities. Where there are threats of serious or irreversible damage, lack of full scientific certainty shall not be used as a reason for postponing cost-effective measures to prevent environmental degradation.

Seeing beyond environmental protection, a definition such as the following reflects what we believe is a typical way of understanding this principle:

> The precautionary principle is the ethical principle that if the consequences of an action, especially the use of technology, are subject to scientific uncertainty, then it is better not to carry out the action rather than risk the uncertain, but possibly very negative, consequences.

The key message is that if there is a lack of scientific certainty as to the consequences of an action, then that action should not be carried out.

The problem with this statement is that the meaning of the term "scientific certainty" is not at all clear. As the focus is on the future consequences of the action, there would be no (or at least very few) cases with known outcomes. Hence scientific uncertainty must mean something else – and three natural candidates are:

(i) knowing which *type of* consequences could occur,
(ii) being able to predict the consequences with sufficient accuracy
(iii) having accurate descriptions or estimates of the real risks, interpreting the real risk as the consequences of the action.

If we adopt one of these interpretations, the precautionary principle could be applied either when we do not know the type of consequences that could occur, or we have poor predictions of the consequences, risk descriptions or estimates. As an example, let us think of the issue about starting year-round petroleum activities in the Barents Sea, see Section 1.2.2. In December 2003 the Norwegian government considered whether year-round activities should be allowed for the areas of Lofoten and the Barents Sea, both ecologically vulnerable areas. Then following (i) and using broad categories of consequences, we cannot apply the precautionary principle as we know the type of consequences of this activity. As a result of these operations, some people could be killed, some injured, an oil spill could occur causing damage to the environment, *etc*. Different categories of this damage could be defined. Hence by grouping categories and types of consequences the possible lack of scientific certainty is "eliminated".

However, in this case, many biologists would say that there is some lack of knowledge as to what the consequences for the environment will be, given an oil spill. This lack of scientific certainty could be classified as fairly small, but that would be a value statement and people and parties could judge this differently. The point is that there is some scientific uncertainty about the consequences of an oil spill. But is this lack of scientific certainty of a different kind than uncertainty related to what will be the outcome of the oil spill? Consider the consequences of an oil spill on fish species, and let X denote the recovery time for the population of concern, with X being infinity if the population does not recover. Then there is scientific certainty according to criterion (ii) if there is scientific consensus about a function (model) f such that X equals $f(Z_1, Z_2 \ldots)$ with high confidence, where Z_1, $Z_2 \ldots$ are some underlying factors influencing X. Such factors could relate to the possible occurrence of a blowout, the amount and distribution of the oil spilled on the sea surface, the mechanisms of dispersion and degradation of oil components, and the exposure and effect on the fish species. For selected values of the Zs, we can use f to predict the consequences X. The precautionary principle applies when it is difficult to establish such a function f – the scientific discipline has not sufficient knowledge for obtaining "scientific certainty" on how the high level performance, in this case measured by X, is influenced by the underlying factors. Models may exist, but they are not broadly accepted in the scientific community.

Scientific consensus in this sense does not mean that the consequences (X) can be predicted with accuracy, when not conditioned on the Zs. Unconditionally, the consequences (X) are uncertain, and this uncertainty is defined by the uncertainties of the factors Z.

To study the criterion (iii), suppose that p represents the "real" risk, quantified by the probability distribution of X, and let p^* be an estimate of p derived from a detailed risk analysis of the activity. Since the uncertainties in this estimate are considered large, relative to the real p, the precautionary principle may be applied following criterion (iii). We see that using (i), (ii) or (iii), we may arrive at different conclusions. In this section we discuss this issue in more detail, specifically addressing the cases (ii) and (iii). The case (iii) is based on some underlying thinking that a real risk exists, but what is this real risk? Other perspectives on risk exist, and how would the understanding of the precautionary principle depend on the perspective assumed? In particular we look closer at a perspective which defines risk as the combination of possible consequences and associated uncertainties, see Section 2.1.

Our conclusions can be summarised as follows: the precautionary principle is a useful concept, with reference to situations in which there is a lack of understanding of how the consequences of the activity being studied are influenced by the underlying factors. In addition the concept "cautionary principle" is important, which says that in the face of uncertainty, *caution* should be a ruling principle. The level of caution and precaution is primarily a management issue, not science.

2.3.1 Discussion of the Meaning and Use of the Precautionary Principle

Having noted the differences in the prevailing perspectives on risk, it is obvious that any discussion of the precautionary principle must make it clear which perspective is taken as the basis. If we mix all the different perspectives together, this will give a rather meaningless analysis, in our view, as the definition and use of the precautionary principle would depend on the perspective. In the introduction we looked briefly at the traditional classical approach to risk. We will return to this perspective, but first we will address the case when probability is used as a subjective measure of uncertainty.

Probability Used as a Subjective Measure of Uncertainty
For this perspective, there is no reference to an objective, real risk, and hence the interpretation iii) in the introduction section does not apply. Probability is here used as a measure of uncertainty as seen through the eyes of the assessors. Consequently, we can restrict attention to the interpretation ii); the criterion i) is discussed in Section 1. Returning to the Barents Sea example, the issue of scientific certainty is related to our ability to determine a function f such that X, the recovery time for the population of fishes, equals $f(Z_1, Z_2\ldots)$ with high confidence, for some underlying factors $Z_1, Z_2\ldots$.

Performing a risk analysis according to this risk perspective, we assign probabilities $P(X<x|K)$, where K is the background information. A lack of scientific certainty as described above is included in the background information. If the assignment is based on the use of a model linking X and some underlying factors Z, we have a lack of scientific certainty if the assignment is based on a model f^* which is not accepted as a good description of the real world.

In practice there will always be some degree of lack of scientific certainty. Hence the question of evaluating this degree is in order. How important is the lack of scientific certainty? How accurate does the model f^* need to be? How can we measure its accuracy?

There are no clear answers to these questions. Different people and parties would judge these issues differently. There are no sharp limits stating that a specific level is not acceptable and that the precautionary principle should apply.

Hence referring to the precautionary principle implies a judgement, expressing the view that we find the lack of scientific certainty *i.e.*, the lack of knowledge, related to how the consequences of the activity are influenced by the underlying factors, to be so significant that the activity should not be carried out. The risk analysis results, producing predictions and uncertainty assessments, provide input to such a judgement.

Applying a broad social science perspective on risk as described in Section 2.1, we then analyse and describe the lay perception of the possible consequences and the associated uncertainties, and this provides a basis for the appropriate management level to decide whether the combination of possible consequences (outcomes) and uncertainties, with all its attributes, suggests a high level of risk. The layman's perception of risk may influence the decision-maker and his/her attitude to the importance of various aspects of the risk picture. This applies in particular to the weights put on the lack of understanding of how the consequences of the activity are influenced by the underlying factors. Hence the lay perception of risk may also affect the application of the precautionary principle.

The concept of uncertainty is the key to understanding the precautionary principle. In the following we address some common ways of structuring and handling uncertainty, and we relate these to the discussion of the meaning and use of the precautionary principle.

A distinction is often made between uncertainty and ignorance. The latter refers to a lack of awareness of factors influencing the issue (HSE 2002). For example, we may have identified a list of possible types of events leading to a major accident. However, some types of events could have been ignored as we are not aware of these or as a means of simplifying the analysis. Adopting a perspective on risk wherein risk is the combination of possible consequences (outcomes) and uncertainties, expressed by subjective probabilities, such a distinction is not critical. The lack of awareness is an element of the uncertainty *i.e.*, the lack of knowledge. If we are not aware of important factors, we cannot establish an accurate model f.

HSE (2001a) refers to three manifestations of uncertainty:

- **Knowledge uncertainty** – This arises when knowledge is represented by data based on sparse statistics or subject to random errors in experiments. There are established techniques for representing this kind of uncertainty, for example confidence limits.
- **Modelling uncertainty** – This concerns the validity of the way chosen to represent in mathematical terms, or in an analogue fashion, the process giving rise to the risks.
- **Limited predictability or unpredictability** – There are limits to the predictability of phenomena when the outcomes are very sensitive to the assumed initial conditions. Systems that begin in the same nominal state do not end up in the same final state. Any inaccuracy in determining the actual initial state will limit our ability to predict the future and in some cases the system's behaviour will become unpredictable.

Some comments related to these concepts are in place. Adopting a perspective on risk, that risk is the combination of possible consequences (outcomes) and uncertainties, there is only one type of uncertainty, and that stems from lack of knowledge related to what the outcome will be. All uncertainties are "knowledge uncertainties". Let us consider an example. A large population of units is imported to country A from country B. A sample of size n is collected to check the quality of the units. Let v be the proportion of failed units in the population and v^* the proportion of failed units in the sample. Clearly the sample provides information and knowledge about v, the total population failure rate, but we do not have full certainty. However, if n is large, we can bound the error $|v^* - v|$ using probability statements, for example expressing a 95% probability that the normalised error (the error divided by the standard deviation) is bounded by a number d. Confidence intervals are not used when adopting subjective probabilities. Being able to control the error term is a way of saying that we have scientific certainty. Experts would agree – there is scientific consensus. Hence for a problem related to sampling from large populations, it seems that the use of the precautionary principle is not relevant. This is not the case, however. We may have an accurate estimate or a prediction of v, but there could be lack of scientific certainty about the consequences of a failed unit – think for example of a unit as a piece of meat, for which there are a number of possible consequences subject to large scientific uncertainties.

Modelling uncertainty does not exist in a context based on subjective probabilities. We assign a probability $P(A|K)$, for an event A, and the models are a part of the background information K. Of course, we need to address the accuracy of the models as discussed above. In risk analysis we use sufficiently accurate models, simplifying the real world. As stated above; if the assignment is based on the use of a model linking X and some underlying factors Z, we have a lack of scientific certainty if the assignment is based on a model f^* which is not accepted as a good description of the real world.

It is also common to distinguish between knowledge uncertainty (epistemic uncertainty) and aleatory uncertainty (stochastic uncertainty). The latter category refers to variation in populations. Using the above case on import of items as an illustration, the variation is given, for example, by the proportion of failed items in the total population. We prefer to use the term variation instead of aleatory uncertainty in such cases, as variation is in fact the meaning. This variation is a basis for expressing uncertainties about observables.

In practice few phenomena can be predicted with certainty. There are almost always uncertainties present. We make a prediction and address uncertainties. For well defined situations, it may be possible to establish functions $X = f(Z_1, Z_2 \ldots)$, so that X can be predicted from the Zs – we have scientific certainty. However, in many cases such functions can only be established in a theoretical world, far from practical risk analysis modelling. Thus we would have scientific certainty, and no need to apply the precautionary principle. Yet the models used could produce poor predictions.

The Search for Real, Objective Probabilities and Risks

We return to the discussion in the introduction of Section 2.3. We have derived an estimate p^* of the real probability p. Except for situations where it is possible to perform sampling of a large number of similar items, the estimate would be subject to large uncertainties. Thus there is a lack of scientific certainty and we may apply the precautionary principle. For the Barents Sea example, the risk estimates would be subject to large uncertainties, and the precautionary principle would therefore be applicable.

Next we discuss the meaning of the different aspects of uncertainty addressed above: knowledge uncertainty, stochastic (aleatory) uncertainty, model uncertainty and limited predictability.

Knowledge uncertainty has already been covered as it is related to the uncertainty of the estimate p^* relative to the real, objective probability p. Confidence intervals could be used to express the uncertainties. Stochastic uncertainty is the variation in the population generating the p. It cannot be reduced by increased knowledge. This is obvious since it is in fact not an uncertainty, but a variation in a given population. Now, what is the population in the Barents Sea example? The probability p expresses the proportion of "experiments" in which the recovery time X exceeds a specific number. The population is a fictional population generated by a thought experiment in which we simulate the activity in the Barents Sea over and over again, with some aspects being stochastic and some other aspects considered a part of the frame conditions of the experiment. For example, the performance of the workers offshore may vary, but the working positions are considered constant. If we lack accurate estimates of this underlying thought-constructed probability, we may apply the precautionary principle. As already noted, this means that for most complex situations in practice we may apply the precautionary principle, if this perspective is adopted, since the estimate would be subject to large uncertainties.

Modelling uncertainty is relevant for this perspective on probabilities and risk, as there is a correct model linking the parameters of the model and the high level probabilities p. A parameter of the model may have the form of a probability or an expected value, for example EZ_i. The uncertainties related to what the correct value of p is, have two main components, the uncertainties in the parameters, and in the model. The model uncertainties are normally too difficult to express, but lead to increased uncertainties in the estimates p^*, and consequently a justification for the use of the precautionary principle.

It is argued in Aven (2003) that this perspective on probability and risk in a way creates uncertainty, not inherent in the object being analysed. The problem is that we need to reflect uncertainty of a mind-constructed quantity – the underlying probability – which does not exist in the real world. Hence the precautionary principle will be given a stronger weight than can be justified from other perspectives.

Concerning the lack of predictability, we refer to the discussion in the previous section.

2.3.2 Conclusions

Among most economists and decision analysts, the theoretical framework for obtaining good decisions is the expected utility theory, based on the use of subjective probabilities. Attention should be on $Eu(X)$, where u is the utility function and X is the outcome. In this framework there is no place for the application of the precautionary principle, as the expected utility is the appropriate guidance for the decision-maker. Uncertainties and the weights put on these uncertainties are properly taken into account using this theory.

However, this is a theory, and it is difficult to apply in practice. People do not behave according to this theory. This is well known, and different alternative frameworks have been suggested. Many economists would refer to the cost-benefit analyses, as the adequate practical tool to guide the decision-makers. By transforming all values to monetary values and calculating expected net present values, $E[NPV]$s, a consistent procedure is obtained for making decisions, which is believed to provide good decisions seen from a societal point of view.

Again, in this framework there is no place for the application of the precautionary principle, as the cost-benefit analysis is the appropriate tool for the decision-maker. However, few people would conclude that the cost-benefit analyses and related tools provide clear answers. They have limitations and are based on a number of assumptions and presumptions, and their use is based not only on scientific knowledge, but also on value judgements involving ethical, strategic and political concerns. The analyses provide support for decision-making, leaving the decision-makers to apply decision processes outside the direct applications of the analyses. It is necessary to see beyond the expected values. This is further discussed in the coming sections of this chapter.

The important question then is how the uncertainties should be taken into account in the decision-making process. The precautionary principle is a way of dealing with the uncertainties. The discussion in the previous sections has demonstrated that the precautionary concept is difficult to understand and use, and depends on the perspective on risk applied.

To us the most meaningful definition of the precautionary principle relates to the lack of understanding of how the consequences of the activity are influenced by the underlying factors, *i.e.*, a version of criterion (ii). If there is a lack of such knowledge, we may decide not to carry out the activity, with references to the use of the precautionary principle. Any reference to being able to accurately measure probabilities should be avoided, as that leads to a meaningless discussion of accuracy in probability estimates. We have to acknowledge that it is not possible to establish science-based criteria for when the precautionary principle should apply. Judging when there is a lack of scientific certainty is a value judgement. In the face of uncertainty, analysts and scientists need to do a good job of expressing the uncertainties, enabling the decision-maker to obtain an informative basis for his or her decision. Based on our experience, there is a large potential for improvement on risk and uncertainty descriptions and communications. Many analysts and scientist have severe problems in dealing with uncertainties, as do many statisticians. Being aware of the different perspectives on risk, and using these in the descriptions and communication, we see as a key element in improving the present situation.

Is there then a need for the concept precautionary principle? Could we not just refer to the possible consequences, the uncertainties and the probabilities *i.e.*, the risks? Well, we need a term for saying that we will not start an activity in the face of large uncertainties and risks, and we will not postpone the implementation of measures because of uncertainties. We may refer to this as a cautionary principle, *cf.* HSE (2001a), but it would be a too broad definition for the precautionary principle. Unfortunately, this kind of broad interpretation of the precautionary principle is often seen in practice. We prefer to restrict the precautionary principle to situations where there is a lack of understanding of how the consequences (outcomes) of the activity are influenced by the underlying factors, and use the concept of caution as the broader principle saying that caution should be the ruling principle in the face of risk. Hence we adopt the cautionary principle when the criterion (ii) is not met *i.e.*, risk is present, and the precautionary principle in the special case described above.

This thinking seems to be consistent with the meaning and use of this principle adopted by HSE in the UK. HSE (2001a, 2003c) adopts the following policy for using the precautionary principle;

Our policy is that the precautionary principle should be invoked where:

- there is good reason, based on empirical evidence or plausible causal hypothesis, to believe that serious harm might occur, even if the likelihood of harm is remote; and
- the scientific information gathered at this stage of consequences and likelihood reveals such uncertainty that it is impossible to evaluate the conjectured outcomes with sufficient confidence to move to the next stages of the risk assessment process.

An essential point here is that the precautionary principle is linked to outcomes and not the risks.

2.4 The Meaning and Use of Expected Values in Risk Management

As the future real *NPV* is an unknown quantity at the time of the planning of an investment project, a related performance measure must be used. In practice this is the *E*[*NPV*]. But is *E*[*NPV*] an appropriate performance measure? Well, if the *E*[*NPV*] is approximately equal to the real *NPV* *i.e.*, *E*[*NPV*] produces accurate predictions of the real *NPV*, the answer should be a yes. In the case of an accident with great losses, it is obvious that the real *NPV* can be quite different from the *E*[*NPV*]. However, this does not matter when having a portfolio perspective. For the company, the return and economic risk for the project itself is of course of interest, but more important is the effect this project will have on the return and economic risk for the company's portfolio as a whole. This follows from the portfolio theory, see Section 2.2. From this theory we see that it is a reasonable approach for the company to attach importance to *E*[*NPV*] and in general expected values for evaluation of the performance of alternatives, in combination with focus on systematic risk. Systematic risk could give large outcome deviations from the expected values. Hence analysis of this risk, for example sensitivity analysis, is required to support decision-making.

There are, however, some additional problems that we have not yet included in the argumentation, which makes evaluation based on expected values somewhat more complicated:

(a) We have restricted attention to production values – the values of lives and the environment have not been incorporated.

(b) We cannot in practice ignore the specific company related risk. Corporate procedures for investment and management could result in large outcome deviations from the expected values. And there are large uncertainties associated with the consequences of an accident – there is a potential for substantial losses.

(c) In an evaluation, we assign probabilities and compute expected values based on a number of assumptions and presuppositions.

In the following we will discuss these problems in more detail.

(a) The Values of Accidents and Lives

The portfolio theory is based on the possibility of transforming all values to one unit, the production value. From a business perspective, moreover, companies may argue that this is the only relevant value. All relevant values should be transformed to this unit. This means that the expected costs of accidents and lives should be incorporated in the evaluations.

But what is the production (economic) value of a life? For most people it is infinite, and very few of us would not be willing to give our life for a certain amount of money. We say that a life has a value in itself, but you may of course accept a risk in return for certain monetary or other benefits. And for the company, this is the way of thinking – the balance of costs and risk. The challenge however, is how to achieve this balance. What are reasonable numbers for the company to use in putting a value on a life in itself? Obviously there are no correct answers, as it is a managerial and strategic issue. High values may be used if it can be justified that this would produce high performance levels, in terms of both safety and production. The issue becomes a problem of the type c).

(b) and (c) Uncertainties in Consequences and Limitations of Calculation Methods

It follows from the portfolio theory that we can ignore specific company related risk. However, in practice we can not ignore this risk because we have corporate procedures in, for example, risk management, and the results of accidents could be large also in a corporate perspective. From time to time we experience accidents that give the company a poor image, with potentially wide-reaching results in terms of market values. And, since the uncertainties in the consequences are so large, the assumptions and suppositions made may greatly influence the results.

To see this more clearly, note that all statistical expected values are conditioned on the background information. In mathematical terms this is written as $E[X|K]$, where X is an observable quantity and K is the background information. The background information covers *inter alia* historical system performance data, system performance characteristics and knowledge about the phenomena in question. Assumptions and presuppositions are an important part of this information and knowledge. We may assume for example in an accident risk analysis that no major changes in the safety regulations will take place for the time period considered, the plant will be built as planned, the capacity of an emergency preparedness system will be so and so, equipment of a certain type will be used *etc.* These assumptions can be viewed as frame conditions of the analysis, and the produced probabilities must always be seen in relation to these conditions. A result of this is that a truly objective expected value does not exist. There could be different values, and different analysts arrive at different values depending on the assumptions and presuppositions made in the project. The differences could be substantial. Expected values should therefore be interpreted with care, as they do not necessarily provide good predictions of the values X.

Consequently, uncertainty needs to be considered, beyond the expected values, which means that the principles of precaution and robustness have a role to play. Furthermore, risk aversion may be justified. The point is that we put more weight on possible negative outcomes than the expected values support. Many companies seem in principle to be in favour of a risk neutral strategy for guiding their decisions, but in practice it turns out that they are often risk averse. The justification is partly based on the above arguments (a)–(c). In the case of a large accident, the possible total consequences could be quite extreme – the total loss for the company in a short- and long-term perspective is likely to be high due to loss of production, penalties, loss of reputation, changes in the regulation regimes, *etc*. The overall loss is difficult to quantify – the uncertainties are large – and it is seldom done in practice, but the overall conclusion is that investments in safety are required. The expected value is not the only basis for this conclusion.

An Example from the Offshore Oil and Gas Industry

We consider the following example in order to illustrate the implications of using expected values. A riser platform is installed with a bridge connection to a gas production platform. On the riser platform, there are two incoming gas pipelines and one outgoing gas pipeline. The pipelines are all large diameter, 36 inch and above. The decision problem is whether or not to install a subsea isolation valve (SSIV) on the export pipeline.

We assume that the analyst has specified an annual frequency of $1 \cdot 10^{-4}$ per year for ignited pipeline or riser failures, *i.e.*, the computed expected number of failures for a one year period is $1 \cdot 10^{-4}$, which is the same as saying that there is a probability of $1 \cdot 10^{-4}$ for a failure event to occur during one year. In the case of an accident, the SSIV will dramatically reduce the duration of the fire, and hence damage to equipment and exposure of personnel.

Let us assume that the computed expected number of fatalities without SSIV is 5, given pipeline/riser failure, and 0.5 with SSIV installed. Let us further assume that the expected damage cost without SSIV is 800 million NOK, given pipeline/-riser failure, and 200 million NOK with SSIV installed. When there is no SSIV installed, the riser platform will have to be rebuilt completely, which is estimated to take 2 years, during which time there is no gas delivery at all. This corresponds to an expected loss of income of 40000 million NOK. With SSIV installed, the expected loss of income is 8000 million NOK.

The expected investment cost is taken as 75 million NOK, and the annual expected cost for inspection and maintenance is 2 million NOK. In the calculations of the expected net present value, 10% interest is used. All monetary values are calculated without taking inflation into account.

The total expected net present value of costs related to the valve is 93.9 million NOK, with annual maintenance costs over 30 years. The annual expected saving (*i.e.* reduced expected damage cost and reduced expected lost income) is 3.26 million NOK, and the expected net present value over 30 years is 30.7 million NOK. This implies that the expected net present value of the valve installation is a cost of 63.2 million NOK.

The expected number of averted fatalities per year is $4.5 \cdot 10^{-4}$ fatalities. Summed over 30 years (without depreciation of lives), this gives an expected value of averted fatalities equal to 0.0135.

Thus, the expected net present value of the costs per averted statistical life lost is 4675 million NOK, and a cursory evaluation of such a value would conclude that the cost is in gross disproportion to the benefit.

But let us examine the results more closely.

It should be noted that if the frequency of ignited failure is 10 times higher, 10^{-3} per year, the expected net present value of the reduced costs becomes 307 million NOK (instead of 30.7 million NOK). This means that the valve actually represents an expected cost saving. In this case, the conclusion based on expected values, should clearly be to install the valve.

The first observation is that the expected net present value of the reduced costs is strongly dependent on the analyst's assignment of the annual frequency for pipeline or riser failures. As discussed above, item (c), we need to see the values produced in the risk analysis in view of the assumptions made in the analysis, the limitations of the analyses *etc.* We should be careful in drawing conclusions based only on the calculated numbers. Sensitivity analyses should always be a part of the decision basis provided.

If we return to the base case values, the probability of experiencing a pipeline or riser failure near the platform is 0.3%, *i.e.*, the scenario is very unlikely. There is a 99.7% probability that there will never be any need for the SSIV, and its installation is just a loss, without any possibility of covering any costs.

But with a small probability, 0.3%, a highly positive scenario will occur. An ignited leak occurs, but the duration of the fire is limited to a few minutes, due to the valve cutting off the gas supply. There are still some consequences; the expected number of fatalities is 0.5, expected damage cost 200 million NOK, and expected lost income of some months, equivalent to 8000 million NOK. These are quite considerable consequences, but they would be much worse if an SSIV was not installed. The expected savings in this case are 4.5 fatalities, 600 million NOK damage cost, and 32000 million NOK in lost income. Note that, in the above calculations, we have disregarded the probability that the SSIV will not work when needed (the error introduced by this simplification is small as the assigned probability of a SSIV failure is small).

If we focus on the economy, there is a probability of 99.7% of a 63 million NOK loss (in expected net present value), and a probability of 0.3% of 32 600 million NOK reduced damage cost (in expected net present value) in a year with a pipeline/riser failure. The expected *NPV*, based on these conditions, becomes 63.1 million NOK. For the installation in question, the expected net present value of 63.1 million NOK is not very informative: either the scenario occurs, with an enormous cost saving (and reduced fatalities) or it does not occur, and there are only costs involved.

From the portfolio theory and a corporate risk point of view, it is still a reasonable approach to use statistical expected values as a tool for evaluating the performance of this project. But, as discussed above, we should not perform mechanical decision-making based on the expected value calculations. We need to take into account the above factors. The conclusion then becomes an overall strategic and political one, rather than one determined by the safety discipline.

2.5 Uncertainty Handling (in Different Project Phases)

We return to the discussion in Section 1.2 of risk handling in different phases. To what extent is the portfolio theory and economic cash flow analyses providing guidance on how to make decisions in projects? To what extent can we ignore the unsystematic risks in project management? To what extent is the use of expected values relevant and appropriate for steering project performance measures, such as production figures, revenues and number of fatalities? What is added by the use of uncertainty and safety management? What are the key factors justifying uncertainty management and safety management? To what extent are the levels of uncertainty and manageability important?

These were some of the main issues raised and in this section we will discuss these issues. Some of our main conclusions can be summarised as follows:

- It is essential to make a sharp distinction between expected values determined at the point of decision-making and the real observations (outcomes). The expected values are to varying degree able to predict the future observations. Uncertainty and safety management are justified by reference to these observations and not the expected values alone.
- Proper uncertainty management and safety management seek to produce more desirable outcomes, by providing insights into the uncertainties of the future possible consequences of a decision.
- Any decision rule, such as the expected *NPV* with a risk-adjusted discount rate, should be supplemented with uncertainty assessments to reveal the potential for uncertainty and safety management in later phases.

The portfolio theory is a theory, and in practice it does not fully apply, see the discussion of Section 2.3:

(1) Expected values should be used with care when an activity involves a possibility of large accidents. Such accidents have a minor effect on expected values, due to their small probabilities, but if they occur they can result in consequences that are not outweighed by other projects in the portfolio.

(2) Assessments of uncertainties are difficult and the probability assignments are based on a number of assumptions and suppositions, and will depend on the assessors' judgements. The expected values computed are not objective numbers.

(3) The specific company risk cannot be ignored because there are corporate procedures in, for example, risk management.

(4) Large accidents most often involve consequences that are difficult to transform into monetary values, and the expected *NPV* can give limited information about the consequences exceeding the strict economic values. What is the value of a life and the environment? How should the company demonstrate for example that a life has a value in itself?

Hence, uncertainty needs to be considered beyond the expected values. The important question then is how the uncertainties should be reflected in the decision-making process, and this is the issue discussed in the rest of this section.

If Y_1, Y_2, … represent future quantities of interest for the decision-maker, such as cash flows, the *NPV*, the number of fatalities, the amount of a toxic substance discharged to sea *etc*., and Y is the vector of these Y_is, we need to distinguish between the expected value EY and the observations Y. We also need to take the time aspect into account. We are to make a decision at time s, say, that has consequences for a future time interval J. Hence we can write $EY = E_s[Y(J)]$, indicating that the expected value is taken at time s, and relates to Y for the time interval J. At time s we have to choose among a set of decision alternatives d_1, d_2, …, and hence we can write:

$$EY = E_s[Y(J) \mid d, K] \tag{2.2}$$

to show that the expected value of Y is given a decision d and the total background information (assumptions and suppositions) K at time s. We look for a decision alternative d that gives the best outcome $Y(J)$. As $Y(J)$ is unknown at the time of the decision, we need to take the uncertainties into account.

It is clear from the discussion of the limitations of the expected values and the Equation 2.2 above for EY, that the expected value could deviate strongly from the observation $Y(J)$. And the reason for this could be factors defined as unsystematic risks. To show the dependency of the expected value of such factors, we may write:

$$Y = g(Z)$$

where $Z = (Z_1, Z_2, \ldots)$ is a vector of factors influencing the quantity Y. Examples of such factors are the narrow pressure margin and the unknown well stream associated with the projects presented in Section 1.2.4. When computing the expectation EY, the values of Z must either be fixed and included in the assumptions as a condition, or the uncertainties should be reflected in the probabilistic analysis. In the former case, an optimistic value is typically assumed, corresponding to a situation where a specific problem will be solved, see the example of Section 1.2.4; the pressure margin is larger than a or the well stream will not differ substantially from the well stream of the primary reservoir. In the latter case more realistic scenarios may be used, but even in this case, aspects of uncertainty are often ignored, as the analysis is always based on some simplifying assumptions.

It follows that a decision to choose between two decision alternatives d_1 and d_2, should not be based on comparisons of the expected values $E[Y(J) \mid d_i, K]$ alone; specific consideration should also be given to the unsystematic risks and uncertainties. The extent to which favourable outcomes of Y can be obtained by proper uncertainty and safety management, must be taken into account in addition to the information gained by computing the expected values. Hence for project I presented in Section 1.2.4, the blowout risk needs special attention, and it would not be sufficient to summarise the blowout risk in one probability number expressing one analyst group's assignment of this probability. The information value of the assigned probability is far less than a comprehensive uncertainty assessment of the possible occurrence of a blowout.

More information about the basis for the uncertainty assessment would strengthen the decision basis. Aspects to consider for a more comprehensive uncertainty assessment include, for example, the composition of the expert group, whether the experts represent the best available information and whether a more detailed analysis would reduce the uncertainty.

In addition to a consideration of the uncertainties and the likelihood of a blowout, information about, for example, the expected geographical dispersion of an oil spill would strengthen the decision basis. In Renn and Klinke (2002) and Kristensen and Aven (2005) a classification scheme with features for a more comprehensive description of consequences is presented, and this consequences classification will be used in the framework presented in Chapter 3.

Another aspect we find important is the manageability of the risk. Some risks are more manageable than others, meaning that the potential for reducing the risk is larger for some risks compared to others. In the example above the process facility risk may be more manageable compared to the blowout risk. The blowout risk is mainly due to difficult pressure conditions in the reservoir, and physical quantities such as reservoir pressures are generally more difficult to affect than, for example, equipment characteristics.

An assessment of the manageability of risk would include some kind of cost-benefit analyses or cost-effectiveness assessment, measuring for example the expected cost per expected saved life. In many cases, such analyses provide sufficient decision support. Other aspects that describe manageability are presented in the framework in Chapter 3.

To illustrate some of these issues we return to the decision problem introduced in Section 1.2.4.

Project Analysis Example Continued

Let us say that the expected *NPV* of the two projects in the analyses, EY_1, are assigned to be 50 million USD and 45 million USD for project I and II, respectively. In addition to EY_1, assessments of the potential consequences, also non-economic, are performed as indicated above addressing, for example, geographical dispersion of an oil spill.

Consider project II in more detail. The cost resulting from process facility problems is one of the performance measures considered relevant for project II. Let Y_2 represent this cost. Assume that a substantial difference between the well stream from the satellite field and the primary reservoir is considered unlikely by experts, and in the magnitude of 1%. Further assume that the cost due to problems with the process facility caused by a substantially different well stream is considered to be in the magnitude of 8 million USD. That is, the expected cost due to process facility problems, EY_2, is 0.08 million USD, and with a probability of 1%, Y_2 is considered to be about 8 million USD.

But can more information be provided to support the assignment of a probability of 1% of a cost of 8 million USD? Yes, of course. For example; knowledge about the expert group used in the analysis would affect the decision-makers' confidence in the assigned values. If the group does not include personnel involved in the design of the original process facility, the result would be reduced confidence in the analysis.

Further strengthening of the decision basis is obtained by considering the manageability of Y_2. The cost due to process facility problems was assessed to be in the magnitude of 8 million USD. Assume that the project plan allows for performing modifications to the process facility, and the relevant personnel are available. Then the costs could be considered much smaller. In the case of a project plan not reflecting the possibility of some upgrading, the costs could be larger than 8 million USD. Thus, the magnitude of Y_2 where there is a large difference between the fields depends mainly on the project planning. Information about such issues will be of value to the decision-maker when considering project II.

The expected *NPV* for the two projects indicates that project I is the most beneficial project. However, more thorough assessments of the consequences, the uncertainty, and the manageability of the projects also influence the decision. For project II the cost due to process facility problems, Y_2, resulting from a large difference between the well stream from the satellite field and the well stream from the primary reservoir is of concern. An assessment of the manageability of this cost shows that large values of Y_2 can be avoided by upgrading the process facility in the case where the satellite field well stream differs substantially from that of the primary reservoir. Such an evaluation should also be given weight when choosing between the two projects. Hence the final assessment of the projects may differ from that indicated by the expected *NPV* analyses alone.

2.6 Risk Acceptance and Decision-making

Safety regulation in the offshore oil and gas industry is largely goal-oriented, *i.e.*, high level performance measures need to be specified and various types of analyses conducted to identify the best possible arrangements and measures according to these performance measures. There has been a significant trend internationally in this direction for more than ten years. On the other hand, there are different approaches taken in order to implement this common objective, if worldwide regulatory regimes are considered.

Whereas the objective may seem simple as a principle, there are certainly some challenges to be faced in the implementation of the principle. One of the main challenges is related to the use of pre-determined quantitative risk acceptance criteria, expressed as upper limits of acceptable risk. Note that in the following, when using the term "risk acceptance criteria", we always have in mind such upper limits. Now, should we use such criteria before any analysis of the systems is conducted? The traditional text-book answer, which is also the prevailing answer to this question in the Norwegian oil and gas industry, is yes. First come the criteria, then the analysis to see if these criteria are met and, according to the assessment results, the need for risk reducing measures is determined. Such an approach is intuitively appealing, but a closer look reveals several problems, of which the following two are the most important;

1. The introduction of pre-determined criteria may give the wrong focus – meeting these criteria rather than obtaining overall good and cost-effective solutions and measures.

2. The risk analyses – the tools used to check whether the criteria are met – are not generally sufficiently accurate to permit such a mechanical use of criteria.

Item 1 is the main point. The adherence to a mechanical use of risk acceptance criteria does not provide a good structure for management of risk to personnel, environment or assets. This is clearly demonstrated for environmental risk. Acceptability of operations with respect to environmental risk is typically decided on the results of a political process and following this process, risk acceptance is not an issue and risk acceptance criteria do not have an important role to play. Risk acceptance criteria have been required by Norwegian authorities for more than 10 years, but almost never have such criteria led to improvement from an environmental point of view.

These issues will be discussed in more detail below. The point here is that there are good reasons to look at other regimes and discuss these against the one based on risk acceptance criteria. The ALARP principle as adopted in the UK sector represents such an alternative. This principle means that the risk should be reduced to a level which is as low as reasonably practicable. Identified improvements (risk reducing measures) should be implemented as a base case, unless it can be demonstrated that the benefits are grossly disproportionate to the costs and operational restrictions. This principle is normally applied together with a limit for intolerable risk and a limit for negligible risk. The interval between these two limits is often called the ALARP region.

In Norway there has recently been a growing focus on the use of risk acceptance criteria. Many risk analysis experts and others are sceptical about the prevailing regime, which applies such criteria more extensively than, for example, corresponding UK practice. The Norwegian Petroleum Directorate (from 1 January 2004 the Petroleum Safety Authority), has up to now regarded such criteria as a cornerstone of the safety legislation regime and has been a driving force for the use of these criteria in the petroleum industry, to control risk related to humans, the environment and economic values. However, the Petroleum Safety Authority has recently raised critical questions about the use of risk acceptance criteria in the industry. The regulations emphasise risk reduction processes but the current focus on risk acceptance criteria has resulted in reduced attention to these processes.

Just as important as the authority requirements is the practice of these requirements by industry and authorities. Throughout this section we will therefore consider also how the practice is carried out by companies and authorities.

It should be noted that except for risk to the environment, there is no exposure of the public from offshore installations; the exposure is limited to employees and the employers' installations.

There is in our view a need to demonstrate that a proper framework for the use of risk analysis can be defined without basing it on risk acceptance criteria. Such a framework would be based on the ALARP principle, but we will not immediately apply the common implementation procedures as seen for example in UK, as we see the need to rethink some of the basic elements of such a framework. We also have to see the framework in relation to the Norwegian safety legislation in general.

The Norwegian offshore oil and gas industry has in many respects been a pioneer in the safety area and the experiences gained should be of interest also outside Norway and other industries. Compared to other regulation regimes, the Norwegian regime has emphasised the use of risk acceptance criteria, and the discussion in the section has to be seen in relation to the Norwegian experience of using such criteria. We believe that we can do better if cost-effectiveness (in a wide sense) is the guiding principle rather than adoption of pre-defined risk acceptance criteria. An essential element in the discussion is the link between political decisions on acceptance and the operator's need to define risk acceptance criteria. A key argument is that if the risk of an activity is judged to be high, the activity is put on the political agenda, and a political decision is made on acceptance, where a proper balance is made between different benefits and burdens. No risk acceptance criteria are introduced. And, given political acceptance, the operators' task is to "optimise" and that should be done without constraints in the form of risk acceptance criteria.

The section is organised as follows. In Section 2.6.1 we summarise the basic elements of the Norwegian risk analysis regime and in Section 2.6.2 we review the common practice of the ALARP principle. In Section 2.6.3 we present and discuss a regime that is not based on the use of risk acceptance criteria at all. Examples are used to illustrate our ideas. In Section 2.6.4 we discuss some of the most common objections against our way of thinking, and finally, in Section 2.6.5, we set out our conclusions. For an in-depth discussion on the ethical justification of the use of risk acceptance criteria, see Section 2.7.

2.6.1 The Present Risk Analysis Regime for the Activities on the Norwegian Continental Shelf

The Norwegian safety regime reflects the basic principle of the licensees' full responsibility for ensuring that the petroleum activity is carried out in compliance with the conditions laid down in the legislation. Since 1985, the safety regime has been founded on internal control, meaning that the authorities' supervisory activities are aimed at ensuring that the management systems of the licensees are catering adequately for the safety and working environment aspects in their activities.

The initial petroleum legislation from the 1970s was technically oriented with detailed and prescriptive requirements for both safety and technical solutions. The authorities with the Norwegian Petroleum Directorate (NPD) in a key role have gradually changed the legislation so that it now has a functional goal orientation.

The NPD regulatory guidelines for concept safety evaluation (CSE) studies were introduced in 1980. The guidelines introduced a quantified cut-off criterion related to the impairment frequency for nine types of accidents that could be disregarded in further evaluation processes, the so-called 10^{-4} criterion, *i.e.*, a maximum probability of 10^{-4} per year for each accident type. These guidelines contributed in a positive manner to using formalised techniques for analysis of risk in the industry, and encouraged the industry and authorities to communicate regarding risk and acceptable risk. However it also had some unfortunate effects, as "number crunching" exercises might be seen as diverting attention from the real issues. Too much emphasis was placed on the methodology and the 'magic' 10^{4} target.

New NPD regulations regarding implementation and use of risk analyses came into force in 1990, and new emergency preparedness regulations appeared in 1992.

The 1990 regulations focused on the risk analysis process. The purpose of risk analyses is to provide a basis for making decisions with respect to choice of solutions and risk reducing measures. According to these regulations, it is the operator's responsibility to define safety objectives and risk acceptance criteria. The objectives express an ideal safety level, thereby ensuring that the planning, maintaining and further enhancement of safety in the activities become a dynamic and forward-looking process. Accidental events must be avoided (any actual accidental event is unacceptable). This means that risk is kept as low as reasonably practicable (ALARP), and attempts are made to achieve reduction of risk over time *e.g.*, in view of technological development and experience. The need for risk reducing measures is assessed with reference to the acceptance criteria. The acceptance criteria and the basis for deciding them are to be documented and auditable.

New PSA regulations relating to management in the petroleum activities came into force on 1 January 2002. These regulations state that the operator has a duty to formulate acceptance criteria relating to major accidents and to the environment. Acceptance criteria must be used for evaluation of results from the various risk analyses and shall be given for

(a) personnel on the installation as a whole, and for personnel groups that are particularly exposed to risk
(b) loss of main safety functions
(c) pollution from installation.

In order to fulfil the requirements and acceptance criteria for major accidents the NORSOK Z–013 standard is recommended.

Some examples of typical risk acceptance criteria used:

- The FAR value should be less than 10 for all personnel on the installation, where the FAR value is defined as the expected number of fatalities per 100 million exposed hours.
- The individual probability that a person is killed in an accident during one year should not exceed 0.1%.

The main characteristic of the present Norwegian system is a relatively "mechanistic" approach to risk analysis and evaluation, implying that efforts are often limited to satisfying the risk acceptance limits, usually with little or no margin.

The result is that there is little or no encouragement for the operating companies to consider if further risk reduction is possible or achievable. When there is little or no margin in an early phase of a development project, this means that later design changes may result in increased risk and exceeding of acceptance limits, often with contractual difficulties between the design contractor and the operating company for the installation in question.

Formally speaking, it may be argued that Norwegian legislation offers the required encouragement for further risk reduction. There is also in the regulations a requirement for an ALARP assessment of risk, in addition to the use of risk acceptance criteria. However, ALARP assessments have already been implemented in the industry to some extent. Where implemented, they too are usually carried out in a mechanistic way. Very often, this process means that possible improvements are identified, but immediately disregarded, based on a cost-benefit (cost-effectiveness) analysis. This analysis is often perfunctory, or very coarse.

In a mechanistic system based on risk acceptance limits, the operator needs to demonstrate to the authorities that the limits have been met. This is often achieved by reference to the risk results, and authority involvement is sometimes rather superficial.

With an ALARP approach, authorities need to be more strongly involved. The ALARP demonstration is more comprehensive than a simple inspection of risk results. For authorities to review an ALARP demonstration, an extensive evaluation process will normally be needed, in order to determine if a sufficiently wide search for alternatives (*e.g.* possible risk reducing measures) was undertaken, and whether arguments relating to gross disproportion are valid. The consequence will be that authorities will need to devote more effort to the task.

This also brings the issue of documentation into focus. Under the Norwegian system, when an operator is ready to commence operation of a new installation, an application for consent to start operation is forwarded to the authorities, based upon a number of studies and assessments, including a number of risk assessments. The authorities give their consent to start operations, if all relevant requirements have been satisfied. No further applications or documents are required until and unless some significant modification is planned after a period of operation.

2.6.2 A Review of the Common Practice of the ALARP Principle

The standard approach when applying the ALARP principle, as practised in the UK sector, for example, is to consider three regions;

1. The risk is so low that it is considered negligible.
2. The risk is so high that it is intolerable.
3. An intermediate level where the ALARP principle applies.

In most cases risk is found in practice to be in region 3 (ALARP region), the ALARP principle is adopted, and an ALARP assessment process is required. This will include a dedicated search for possible risk reducing measures, and a subsequent assessment to determine which of these to implement.

The risk acceptance criteria used in Norway are typically lower than the intolerability limit and higher than the negligible level. But we see a tendency to define risk acceptance criteria which are close to the intolerability levels used on the UK sector.

In the UK, the ALARP principle applies in such a way that the higher the risk, the more employers are expected to spend to reduce it. At high risks, close to the level of intolerability, they are expected to spend up to the point where further expenditure would be grossly disproportionate to the risk; *i.e.*, that costs and/or operational disturbances are excessive in relation to the risk reduction. This is generally considered to be a reasonable approach as higher risks call for greater spending. More money should be spent to save a statistical life if the risk is just below the intolerability level than if the risk is far below this level.

Guidance values are sometimes used, in order to illustrate what values define "gross disproportion". When specifying such numbers, we have to clarify whether the company cost only or societal costs are included, is it before or after tax, with or without insurance compensation, *etc.*

Societal investments in risk reducing measures are sometimes analysed in order to identify the costs spent to avoid loss of a statistical life. Such values may vary from substantially less than 1 million NOK, up to more than 100 million NOK. The societal costs of an average fatality in an accident in Norway have been calculated by SINTEF (SINTEF 1992) as around 25 million NOK.

A typical number for a value of statistical life used in cost-benefit analysis is 1–2 million GBP (HSE 2006, Aven and Vinnem 2005), which corresponds well to 25 million NOK. This number applies to the transport sector. For other areas the numbers are much higher, for example in the offshore UK industry it is common to use 6 million GBP (HSE 2006). This increased number accounts for the potential for multiple fatalities and uncertainty.

It is known that one oil company has a guidance value around 200 million NOK for use in ALARP analysis. A comprehensive ALARP assessment from upgrading of an existing installation is presented in Vinnem *et al.* (1996). Most of the proposed risk reducing measures were determined on the basis of qualitative evaluations and considerations. When quantitative analysis of costs and benefits was finally performed, it was found that among those measures that had been rejected, the measure with lowest cost per averted statistical life lost corresponded to almost 750 million NOK per expected life saved.

The ALARP principle implies what could be referred to as the principle of "reversed onus of proof". This means that the base case is that all identified risk reduction measures should be implemented, unless it can be demonstrated that there is gross disproportion between costs and benefits. To verify ALARP, procedures mainly based on engineering judgements and codes are used, but also traditional cost-benefit analyses and cost-effectiveness analyses. When using such analyses, guidance values as indicated above are often used, to specify what values define "gross disproportion".

The practice of using traditional cost-benefit analyses and cost-effectiveness analyses to verify ALARP has been questioned (Aven and Abrahamsen 2006). The ALARP principle is an example of application of the cautionary principle. Uncertainty should be given strong weight, and the grossly disproportionate criterion is a way of making the principle operational. However, cost-benefit analyses calculating expected net present values ignore the unsystematic risks (uncertainties) and the use of this approach to weight unsystematic risk is therefore meaningless.

Modifications of the traditional cost-benefit analysis are suggested to solve this problem, see EAI (2006) and Hallegatte (2006). In these methods, adjustments are made to either the discount rate or the contribution from cash flows. This latter case could be based on the use of certainty equivalents for uncertain cash flows. Although arguments are provided to support these methods, their rationale can be questioned. There is a significant element of arbitrariness associated with the methods, in particular when seen in relation to the standard given by the expected utility theory.

To explain this in more detail, say that the net present value relates to two years only, and the cash flows are X_0 and X_1. Then an approach based on certainty equivalents means an expected utility approach for the cash flows seen in isolation. The uncertain cash flows are replaced by their certainty equivalents c_0 and c_1, respectively, which means that the uncertain cash flow X_i is compared to having the money c_i with certainty, $i = 0,1$. The specification of such certainty equivalents is not straightforward, see the review of the expected utility theory in Section 2.2.1. However, the important point here is not this specification problem, but the fact that this procedure does not necessarily reflect the decision-maker's preferences. If we ignore the discounting for a second, the utility function of the cash flows X_0 and X_1 is not in general given by the sum of the individual utility functions. By introducing certainty equivalents on a yearly basis, we take uncertainties into account but the way we do it has not been justified.

The alternative approach of adjusting the discount rate seems plausible as the systematic risk is incorporated in the net present value calculations through this procedure. But how large should the adjustment be? There is a rationale for systematic risk adjustment – the CAPM model – but there is no such rationale for unsystematic risk. It will be impossible to find such a rationale, in fact, as the calculations are based on expected cash flows, which ignores the uncertainties. Hence we have to conclude that such an adjustment cannot be justified.

The common procedures for verifying the grossly disproportionate criterion using cost-benefit analysis therefore fail, even if we try to adjust the traditional approach. We should be careful in using an approach which is based on a conflicting perspective, ignoring unsystematic uncertainties.

So what alternative would we then suggest? In our view we have to acknowledge that there is no simple and mechanistic method or procedure for balancing different concerns. When it comes to the use of analyses and theories we have to adopt a pragmatic perspective. We have to acknowledge the limitations of the tools, and use them in a broader process where the results of the analyses are seen as just one part of the information supporting the decision-making. And the results need to be subject to an extensive degree of sensitivity analyses. We refer to Chapter 3.

Under the UK system, the installation cannot be operated until the authorities have accepted the Safety Case, where the ALARP demonstration is one of the main elements. There may be some difference in the approaches taken, in the sense that giving acceptance can be considered as somewhat more active than giving consent. It should be emphasised that risk assessments as part of the safety cases often have the same weaknesses as those conducted under Norwegian legislation, and that they do not reflect operational aspects sufficiently well. This may imply that risk results will not change significantly during the operational period, if no major modifications have been implemented. The ALARP assessment may nevertheless change, if new information has been made available from research, from experienced accidents or incidents or changes in the way performance standards for safety critical systems are fulfilled on the installation in question.

This underlines the fact that an ALARP assessment has no "eternal life"; it is a dynamic process which needs to be reconsidered regularly, in the light of new experience and new data.

In the UK risk intolerability levels are not considered to be instruments of precise control of risk. Compared to the Norwegian system, the UK regime puts stronger emphasis on the ALARP process, reflecting the need to put risk analysis results into a broader context of risk reduction, taking into account the limitations and constraints of the analyses.

2.6.3 A Structure for a Risk Analysis Regime Without the use of Risk Acceptance Criteria

General

Our starting point is a decision-maker facing some decision points in a project. These decision points include problems and opportunities, such as poor HES results, implementation of a risk reduction policy, the use of new technology, choosing a concept for further evaluations, *etc.* Having identified the main decision points, adequate decision alternatives need to be generated and evaluated, relating to whether or not to execute an activity, alternative concepts, design configurations, risk reducing measures, *etc.* Our focus is on situations characterised by a potential for rather large consequences, large associated uncertainties and/or high probabilities of what the consequences will be if the alternatives are in fact being realised, *i.e.*, high risks according to our definition of risk. The consequences and associated uncertainties relate to economic performance, possible accidents leading to loss of lives and/or environmental damage, *etc.* Risk and decision analyses are considered to give valuable decision support in such situations, and according to the present risk analysis regime in Norway, risk acceptance criteria should be used together with the results from these analyses as input to risk evaluation. In this section, however, we will present and discuss an approach where such criteria are not adopted at all. The question is whether such a principle can be justified, and what the pros and cons of such a principle are.

Before presenting a detailed approach for a risk analysis regime without the use of risk acceptance criteria, we will briefly discuss a simple model of the decision process. The model, shown in Figure 2.1, covers the following items:

❶ Stakeholders. The stakeholders are here defined as people, groups, owners, authorities *etc.* that have interest related to the decisions to be taken. Internal stakeholders could be the owner of the installation, other shareholders, the safety manager, unions, the maintenance manager *etc.,* whereas external stakeholders could be the safety authorities (the Norwegian Safety Petroleum Authority, the State Pollution Control Agency, *etc.)*, environmental groups (Green Peace *etc.)*, research institutions, *etc.*

❷ Decision problem and decision alternatives. The starting point for the decision process is a choice between various concepts, design configurations, sequence of safety critical activities, risk reducing measures *etc.*

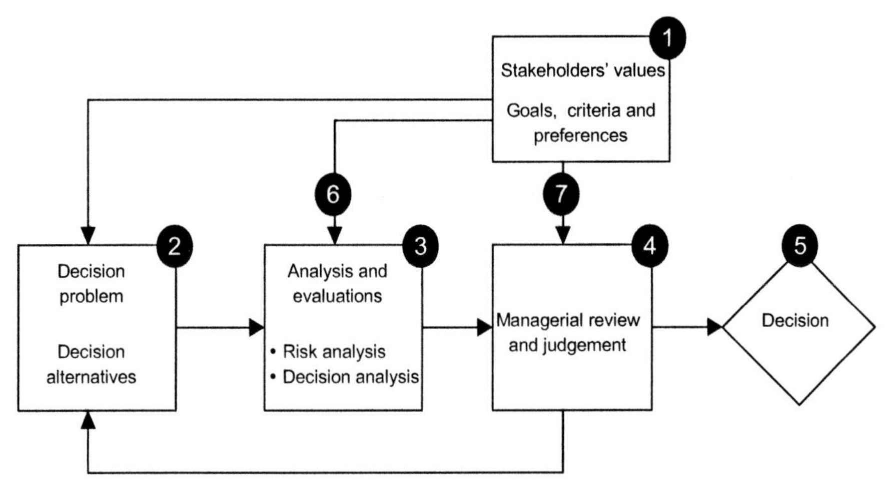

Figure 2.1. Model of the decision-making process (Aven, 2003)

❸ Analysis and evaluation. To evaluate the performance of the alternatives, different types of analyses are conducted, including risk and cost-benefit (cost-effectiveness) analyses. These analyses may, given a set of assumptions and limitations, result in recommendations on which alternative to choose.

❹ Managerial review and judgement. The decision support analyses need to be evaluated in the light of the premises, assumptions and limitations of these analyses. The analyses are based on background information that must be reviewed together with the results of the analyses. Consideration should be given to factors such as

- The decision alternatives being analysed
- The performance measures analysed (to what extent do the performance measures used describe the performance of the alternatives?)
- The fact that the results of the analyses represent judgements and not only facts
- The difficulty of assessing values for burdens and benefits
- The fact that the analysis results apply to models *i.e.*, simplifications of the real world, and not the real world itself. The modelling entails the introduction of a number of assumptions, such as replacing continuous quantities with discrete quantities, extensive simplification of time sequences, *etc*.

In Figure 2.1 we have indicated that the stakeholders may also influence the final decision process ❼ in addition to their stated criteria, preferences and value tradeoffs providing input to the formal analyses ❻.

A More Detailed Structure

To make a risk analysis regime workable without the use of risk acceptance criteria, a procedure such as the following could be appropriate:

1. Perform a crude analysis of the benefits and burdens of the various alternatives addressing attributes related to feasibility, conformance with good practice, economy, strategy considerations, risk, social responsibility, *etc.* The analysis would typically be qualitative and its conclusions summarised in a matrix with performance shown by a simple categorisation system such as Very Positive, Positive, Neutral, Negative, Very negative. From this crude analysis a decision can be made to eliminate some alternatives and include new ones, for further detailing and analysis. Frequently, such crude analyses give the necessary platform for choosing one appropriate alternative.

   When considering a set of possible risk reducing measures, a qualitative evaluation in many cases provides a sufficient basis for identifying which measures to implement, as these measures are in accordance with good engineering or with good operational practice. Also many measures can quickly be eliminated as the qualitative analysis reveals that the burdens are much more dominant than the benefits.
2. From this crude analysis the need for further analyses is determined, to give a better basis for deciding which alternative(s) to choose. This may include various types of analyses of risk.
3. Often the risk analysis focuses on the possibility of loss of lives. Then the risk analysis presents a risk picture related to this consequence, and this risk picture is compared with other relevant activities, analyses and data. From this evaluation, the analysis group has a basis for giving a statement about how they judge the risk. The analysis group does not conclude on whether risk is acceptable or not, as acceptance is related to the alternative considered, with all benefits and burdens associated with it, and not only the risk level.
4. Other types of analyses may be conducted to assess, for example costs, and indices such as expected cost per expected saved statistical life could be computed to provide information about the effectiveness of a risk reducing measure or compare various alternatives. The expected net present value may also be computed when found appropriate. Sensitivity analyses should be performed to see the effects of varying values for statistical lives and other key parameters.

   Often the conclusions are quite straightforward when calculating indices such as the expected cost per number of expected saved lives over the field life and the expected cost per averted ton of oil spill over the field life. If there is no clear conclusion about gross disproportion, then these measures and alternatives are clear candidates for implementation.

   Clearly, if a risk reducing measure has a positive expected net present value it should be implemented. Crude calculations of expected net present values, ignoring difficult judgements about valuation of possible loss of lives and damage to the environment, will often be sufficient to conclude whether this criterion could justify the implementation of a measure. An example is shown by Vinnem *et al.* (1996).

5. An evaluation of other factors such as risk perception and reputation should be carried out whenever relevant, although it may be difficult to describe how these factors would affect the standard indices used in economy and risk analysis to measure performance.
6. A total evaluation of the results of the analyses should be performed, to summarise the pros and cons for the various alternatives, where the constraints and limitations of the analyses are also taken into account.
7. The decision-maker then performs a review and judgement of this decision support and makes a decision.

The essential element in the above decision process is a drive for generating alternatives. Often a base case is defined, but the successful implementation of this regime is that there is a climate for considering possible changes and improvements compared to the base case. If risk to personnel or the environment is considered relatively high, solid arguments will be required not to improve or eliminate the alternative. The difference in costs would have to be grossly disproportionate if no safety improvements were made. If an alternative is chosen with a fairly high risk level, the decision-maker must be able to document the arguments in case of later scrutiny, for example in the event of an accident.

It is essential that the analysis team has the ability to communicate the information from the analyses to the decision-maker, and the decision-maker must understand what the analyses and the analysts express. Compared to the present situation, there is a need for improvements in both these areas. It is also necessary that the results from the analyses are communicated to management at a sufficiently high level. The implications of risk results may sometimes be far reaching, with facets that are non-tangible, and with certain dimensions of a political nature. It is therefore important that risk results should be communicated directly to a high management level, and not filtered through several layers of middle management.

Compared to a regime based on the use of risk acceptance criteria, the above regime could in some cases mean a more direct visualisation of the decision-maker's trade-offs between safety and other aspects, such as costs. Some may think this is appropriate, but it could also be a problem – not all decision-makers would see this as attractive. The use of risk acceptance criteria means an extended level of delegation to lower levels of decision-making.

2.6.4 Cases

In the following we will discuss the use of risk acceptance criteria in the offshore oil and gas industry, using the Norwegian safety regime as the basis. We will argue that the use of risk acceptance criteria is difficult to justify. We will do this by distinguishing between different phases for judgement of risk acceptance. First we look at the early phase of an activity, which is often characterised by high potential for reducing the risk to personnel. Then we consider the execution phase where the risk level has already been judged as tolerable, on an overall level, and the goal is to optimise arrangements, operational plans and measures. As illustrations, let us look at the initial phase of the drilling for oil and gas in the North Sea and the use of helicopters for transporting people to and from offshore installations. Any evaluation of these activities would immediate conclude that they are high risk activities. The benefits of the activities are large, however, and they are therefore accepted. The risk level is considered acceptable or tolerable. This conclusion is established as a result of a political and management process – it is not the result of formal risk assessments using pre-defined risk acceptance criteria. The politicians' and the managers' task is to balance different benefits and burdens, and often this means balancing benefits and risks. If the risk level is high, the benefits of the activity are also often high, and the activity is put on the political agenda. Clearly, the use of pre-defined risk acceptance criteria is not appropriate for situations like this. What is gained by specifying for example a probability of having one or more fatalities during a year of say maximum 0.01? It means less flexibility for making an overall balanced consideration of the various burdens and benefits of the activity. It could even stop the realisation of the activity. And it would lead to the wrong focus, resulting in a more or less unfounded number. What should the proper number be? Clearly, there is no universally appropriate number to be used, as the risk accepted is a function of all the burdens and benefits of the activity. If we want to use risk acceptance criteria, we would need to adopt criteria that make the activity acceptable with respect to risk level. But what is then the benefit, in such circumstances, of using the criteria? Given the established practice (with respect to *inter alia* technology), risk is considered acceptable (tolerable) and the risk calculations showing that risk is below the acceptance criteria are in fact not important – it does not provide decision support. Unfortunately, much of the present use of risk analysis and risk acceptance criteria is of this kind. Instead, risk analysis should be used to identify critical areas and improvement potential and assess the effects of possible safety improvement measures.

Political acceptance may result in a demand for changes and improvements to concepts, designs, plans, *etc*. This could be based on specific requirements related to arrangements and measures, or it could be related to explicit considerations of risk. In the latter case, the risk reductions have to be seen in relation to the cost of implementing the arrangements and measures. Again, a strict adoption of pre-defined limits of acceptable risk levels would reduce the flexibility required to realise activities considered to provide more benefits than burdens.

Now, let us take the case that a decision has been made to develop an offshore oil and gas field. Then this activity is found to be a "tolerable (acceptable) activity", given the boundary conditions (laws and regulations, available technology, *etc*.). The management of the company then needs to "optimise" the concepts, arrangements and measures. The question is, however, how to optimise the use of resources to obtain the best possible concepts, arrangements, operational sequences and measures, balancing the various burdens and benefits, as the activity's risk level is considered tolerable at this level of precision. This means that pre-defined sharp constraints in this optimisation process should be avoided or at least reduced to a minimum. Some examples will be used to illustrate this.

Design of an Offshore Installation

Standard design procedures would in most cases give concepts and arrangements with a risk level which designers and safety professionals find "tolerable", although there could be large potential for improvements. If they do not find it tolerable, they would in most cases not present it as an alternative. And as discussed above, the present level of technology, operational procedures, *etc.* have already been judged acceptable at an earlier stage of the development phase, through the overall political and management processes. What is tolerable or acceptable may be defined by comparing the risk level of the system considered to that with the highest risk level among comparable "approved" systems, or to some specified level of risk. But this number is of minor importance as risk should be substantially lower (since it is an upper limit) than this number.

Further risk analysis is not justified to obtain risk acceptance, but to provide decision support on the effectiveness of possible improvements or changes. The goal should be optimisation of arrangements, plans and measures, in order to obtain the proper balance between costs and reduced risk, as well as other burdens and benefits. Using risk acceptance criteria, the risk analyses act as a verification tool to check compliance or non-compliance. Examples of risk acceptance criteria used in a situation like this are upper limits of FAR-values and *f-N*-curves. These risk indices are all showing some average performance for the whole population of personnel, or major groups of personnel, but this risk is already considered tolerable on an overall level as discussed above. What remains to be controlled is the individual risk – we may refer to this as risk considerations on a second-order level of detail – but this risk is outside the scope of the risk analysis in design. Most operational and organisational factors are optimised in the operational phase, reference is made to the example below concerning helicopter transport.

The same type of reasoning applies to possible accidents leading to environmental damage. The overall environmental risk has already been accepted, when execution of the activity was accepted. Second-order level risk considerations should be an optimisation exercise without unnecessary constraints.

To perform the optimisation of the concept, it is common practice to use predefined requirements, related to safety function impairment frequencies and effectiveness of safety barriers, to specify dimensioning loads, for example.

An example is the requirement for maximum design wave load. If we go back to around 1960, the common design approach for offshore Gulf of Mexico was to design for waves with 25 year return periods. One year in the early 1960s, there were many severe storms, leading to more than a dozen platforms toppling over due to wave overload. It was then decided that dimensioning wave load should be increased to 100 years return period in order to increase the minimum standard. 100 year return periods are still used in the North Sea and other Norwegian waters as the minimum design wave load, without significant damage to the structure. In the last 10–15 years, it has also been required in Norwegian operations that installations should be able to withstand waves with a 10,000 year return period, but then substantial damage to the structure is acceptable.

From a theoretical point of view, our arguments also apply to such requirements – they represent unnecessary constraints and should be avoided. However, for practical work the use of the pre-defined requirements of this type is justified as they simplify the planning process. To perform the optimisation there needs to be some initial characterisations of the performance of the systems of the installation. Care has to be taken, however, to avoid suboptimisation. Over-stringent requirements regimes may limit creativity and the drive to identify the best overall arrangements and measures. We refer to the conclusion section for further comments related to the implementation of such requirements.

Concept Optimisation

Concept optimisation is that point in the development of an offshore installation where the potential for influence from risk assessment is among the highest. It is perhaps also the time when many experts will want to use risk acceptance criteria. But, as we will try to illustrate, it is also the time with the highest potential for achieving extra risk reduction through an ALARP evaluation.

We will discuss two scenarios in relation to concept optimisation:

(a) The concept as initially presented has a FAR value just **below** the acceptance limit.

(b) The concept as initially presented has a FAR value somewhat **above** the acceptance limit.

If the FAR value is below the risk acceptance limit, the risk assessments will normally perform some kind of search for further improvement, but this search is often not very extensive, and is conducted with limited motivation for implementing improvements.

If the FAR value is somewhat above the acceptance limit, there will be a dedicated search for risk reducing measures. Consider for example that a large contribution comes from high exposure of personnel during evacuation. This will often result in more detailed modelling of the escape of personnel from hazardous areas back to the shelter area. More detailed modelling could well bring the average FAR value below the risk acceptance limit, as the modelling often reduces any "conservatism" in the analysis. If not, additional heat shielding on escape ways may be needed. But let us assume that the concept has a real problem for safe escape in certain accident conditions, such that a more fundamental solution is to provide an under-deck escape tunnel with overpressure protection and H0 rating, which for all fires on deck and on sea would be safe to use. Let us assume that the extra cost implied by such an escape tunnel is 20 million NOK.

In the circumstances we have described here, it is likely that in both cases limited improvements (*i.e.* less than installation of an extra escape tunnel) would be chosen, if risk acceptance criteria form the main principle for the control of risk in safety management.

In our view, the only safety management regime which would guarantee serious consideration of the need for fundamental improvement is one in which ALARP is the ruling principle in risk control, without any risk acceptance criteria. With the values indicated, the cost of such an escape tunnel would in most circumstances **not** be considered to be grossly disproportional, especially if it was believed that the concept actually had a problem with respect to the provision of safe escape possibilities. Thus one would expect the escape tunnel to be installed, if it complies with good engineering practice and is decided upon at the concept optimisation stage.

This implies that although the potential for relatively inexpensive improvement of a concept is at the concept optimisation stage, the use of risk acceptance criteria for risk control may be the highest obstacle to realisation of such improvement.

Shuttling Between Installations

We will address two issues:

a) the total risk of fatalities
b) the risk for individuals.

To simplify, we may assume that the risk is proportional to the number of flights n (defined in a certain period of time). Now, should we define acceptable risk limits by saying that the probability of accidents (or fatalities) should be maximum x, and the individual probability of being killed due to shuttling shall not exceed y? Based on our simplification, these probabilities are functions of the number of flights n, meaning that stipulating limitations on the number of flights sets limits for the risks.

Both these issues are closely linked to the manning levels of the installations, and consequently the cost of operating these installations. Thus we need to see the risks in a broader context, which also involves workers' unions and politicians. Given the present level of activity on the Norwegian Continental Shelf, which is in fact approved by the Norwegian parliament, and the present regime for manning the installations, the total risk (a) is *a priori* considered tolerable. A job has to be done, and spreading the risk over a longer period of time in order to reduce the risk level in one particular year does not reduce the accumulated risk, and has no effect. However, in case (b), restrictions on the exposed risk are meaningful, as there could be pressure to expose some people to higher risks than they appreciate. This must naturally be seen in relation to the existing structures for remuneration, but it seems sensible to impose some general restrictions on the number of flights for each person. Whether this is a result of a risk being judged as acceptable or not or the number of flights as too high, is a matter of choice, as the link between the solution n and risk is so clear in this case.

Observe that the specification of such a limit n is not the same as using a pre-determined risk acceptance criterion, as argued against above. The decision to be taken is to determine n and for that purpose a procedure using the principles discussed above, with the generation of alternatives (different n values) and assessment of the associated consequences, may be adopted. This process generates a specific solution, the proper n. The shuttling risk analysis is used as a basis for specifying this n.

It should be noted that restrictions on exposure of personnel to helicopter transport may often lead to increased number of flights, which will inevitably increase the total exposure of helicopter pilots. This could be considered a negative factor, which underlines that there are complex links between the different factors.

Modifications on an Installation in Operation

One frequent argument for using risk acceptance criteria in the operational phase is that such criteria are considered useful for controlling the risks after years with modifications on the installation. The question addressed is then; is the risk still acceptable? Again, our argument would be that if the risk is considered high, changes have been made which means balancing high benefits and increased accident risks, and this would be judgements on a high management level, which should not be constrained by risk acceptance criteria. If the changes are accepted, the ALARP principle should apply. We refer to the example above on design of an offshore installation.

2.6.5 Common Objections to our Approach

Some of the most common objections to our thinking are discussed below:

(1) Risk acceptance criteria are needed to ensure a minimum safety level. Without such criteria we may expect a significant reduction in safety level, and economic concerns can be used to reduce the safety level.

Response: No, this is not the case. The regulations have a number of specific requirements ensuring a minimum safety level. If the calculated risk is substantially above the risk acceptance criteria – the levels of tolerability (we write substantially, as the precision level of risk analysis does not allow "millimetre" considerations) – this would not be consistent with the regulations stating that a high safety level should be established, maintained and further developed. In such a case, we would have no confidence in being able to avoid accidents, and interventions from the authorities are required. In the case of large uncertainties, the regulations state that arrangements should be made reducing the uncertainties. Again this would be an argument for not accepting such risks. Furthermore, such high risks would also mean political involvement, and this would put additional pressure on the operator.

An additional remark is in place. We have to acknowledge that it is the oil companies that specify the risk acceptance criteria and most of the requirements in the Norwegian offshore sector. The whole safety legislation is based on the operator having the full responsibility for its activity, and the regulations allow the operators a large degree of freedom to find adequate solutions and measures. If we do not rely on the oil companies, the regulatory regime must be changed.

(2) Decision criteria will always be needed in order to make decisions. Risk acceptance criteria are just a form of decision criteria.

Response: We do not disagree that some kind of decision criteria will be needed in some circumstances. Consider for example the above illustration on shuttling between installations where some criteria relating to the allowed volume of helicopter flying per person may be needed in some circumstances. However, the mechanical application of pre-defined risk acceptance criteria need not be the answer.

(3) There are so many decisions to be made on a daily basis that having to perform an extensive "process" each time would be too time consuming.

Response: We agree that there are many decisions taken, and that they need to be taken in an efficient manner. Few daily decisions however, make reference to risk acceptance criteria: they are based on other types of decision criteria, specific requirements, qualitative evaluations, *etc*. For the few times that risk acceptance criteria are used in the decision process, a decision process may be substituted, without extensive delay or inefficient use of resources.

(4) Risk acceptance criteria have been used for more than 20 years, and are functioning well.

Response: We think this argument may be challenged. What probably has functioned well is that decisions have been made efficiently, but have they been the right decisions? We think that better decisions (right decisions) may be arrived at through an alternative approach, whereby higher level management becomes more involved in the decision-making process.

(5) A balanced evaluation of burdens and benefits often becomes very complex, because:

- Burdens and benefits are not realised at the same time: some may be delayed or spread out over a long period. Net present value is accepted for income and costs that are measured in monetary values, but not at all for other consequences such as loss of life or environmental effects.
- Burdens and benefits may not affect the same groups of people, or may affect them at different times, thus making "equal risk distribution" an impossibility.
- The evaluation of burdens and benefits will need to be restricted to what the operating company is capable of influencing, which for instance is restricted to working hours, and not off-duty hours.

Response: We agree that a balanced evaluation of burdens and benefits is sometimes very challenging. It becomes neither more nor less challenging through the avoidance of pre-defined risk acceptance criteria in the evaluation process.

(6) The present system of using risk acceptance criteria should be gradually developed away from the present relatively mechanistic into a system of balanced evaluation of burdens and benefits. A gradual development will be better than an abrupt change of approach.

Response: If a gradual transition to a different system can be achieved, that may be the best solution. We do not address the problem of how a transition or change should be implemented in practice.

On the other hand, gradual change from a system that has been mainly unchanged for many years may be difficult to achieve. It may be argued that authorities have tried to promote a gradual change for some time, but without success. Authority representatives have made the comment for some years already, that they sometimes experience use of risk analysis which is virtually contrary to what has been intended. Risk analysis and its results are sometimes interpreted as "evidence" of no need for change. In spite of these comments, no changes have occurred.

(7) Risk acceptance criteria may give the safety professionals the opportunity to at least achieve a minimum of risk reduction, if the management (of a project or an installation) does not give accident prevention sufficiently high priority.

Response: If project or installation management does not give sufficiently high priority to accident prevention, one could argue that it would be better to let that conflict be visible, to be corrected by the workers' representative system, upper level management or authorities. The "solution" provided by risk acceptance criteria in these circumstances is probably only a minimum solution, in the sense that only the least possible risk reduction will be approved.

2.6.6 Conclusions

In the above, we have argued for the need to consider risk as a basis for making decisions under uncertainty. Such considerations, however, must be seen in relation to other burdens and benefits. Care should be shown when using predetermined risk acceptance criteria in order to obtain good arrangements, plans and measures. Pre-defined criteria driving the decisions should in general be replaced by a risk management approach highlighting risk characterisation and evaluation, and a drive for risk reductions and a proper balance between burdens and benefits.

Risk analyses support decision-making on choice of specific concepts, arrangements, measures, procedures, *etc.*, as well as decision criteria. Such decision criteria may have the form of a requirement, for example, the system should have a probability of failure of maximum 1/1000 for a period of one year. Further detailing of this system in a later development phase, could involve risk/reliability/performance analyses to support decision-making, and 1/1000 would be a boundary condition for system performance. Some people may also refer to 1/1000 as a pre-determined risk acceptance criterion. This example illustrates the different levels of criteria used for supporting decision-making, and the need to view the development of criteria and requirements in a time perspective. Above, we have mainly focused on the high level criteria used for the total system and not its many subsystems and components. For the latter it may be more appropriate to apply specific acceptance limits, to facilitate the design and development process, but even for such situations our main approach could be used. Generating alternatives and predicting their burdens and benefits should always, in our view be the ruling paradigm.

We see that there is a hierarchy of goals, criteria and requirements. These can schematically be divided into four categories:

1. overall ideal goals, for example "our goal is to have no accident",
2. risk acceptance criteria (defined as upper limits of acceptable risk) or tolerability limits, managing the accident risk, for example "the individual probability of being killed in an accident shall not exceed 0.1%",

3. requirements related to the performance of safety systems and barriers, such as a reliability requirement for a safety system,
4. requirements related to the specific design and operation of a component or subsystem, for example the gas detection system.

The main message from the discussion in Section 2.6 can be summarised as follows:

- Focus should be on meeting defined overall objectives, which should be formulated using quantities that are observable (such as the number of fatalities, the number of injuries, the occurrence of a specific accidental event, *etc.*). Probabilistic quantities should not be used to express such objectives.
- Safety management is a tool for obtaining confidence in meeting these objectives.
  - Emphasis should be placed on generating alternatives, to be compared with projected performance.
  - Risk acceptance criteria (level 2 above) should not be used.
  - To ease the planning process for optimising arrangements and measures, requirements related to safety systems and barriers may be useful (level 3 above).
- What is acceptable from a safety point of view and what constitutes a defensible safety level, cannot in principle be determined without incorporating all the pros and cons of the alternative, and the decision needs to be taken by personnel with formal responsibility at a sufficiently high level.

2.7 On the Ethical Justification for the Use of Risk Acceptance Criteria

In this section we are concerned about the ethical justification of the use of risk acceptance criteria, and as an illustrative example, let us consider the Norwegian offshore oil and gas activities and the regulations on health, safety and environment (HES) issued by the Norwegian Safety Petroleum Authority (PSA). The regulations include a number of specific requirements related to HES, for example specifying the capacity of the fire walls protecting the living quarters. Most requirements are of a functional nature, saying what to achieve rather than how to achieve it. In addition to such requirements, the PSA regulations require that the operators specify risk acceptance criteria for major accident risk and environmental risk, see Section 2.6.1. By acceptance criteria we mean the upper limit of acceptable risk relating to major accidents and risk relating to the environment. Acceptance criteria must be set for the personnel on the facility as a whole, and for groups of personnel who are particularly risk exposed, pollution from the facility and damage done to third parties. These acceptance criteria are to be used in assessing results from the quantitative risk analyses.

The regulations state that over and above the level given by these requirements, the specific requirements and the risk acceptance criteria, risk must be further reduced as far as possible, *i.e.*, the ALARP principle applies. Hence we may see the specific requirements and the risk acceptance criteria as minimum requirements to be fulfilled by the operators. The justification for these minimum requirements concerning people and the environment is ethical – people and the environment should not be exposed to a risk level exceeding certain limits. Having established such minimum requirements, the authorities' supervision can be quite easily carried out by checking that these requirements are met. Hence the authorities are in a position to decide if the HES level is acceptable or not, depending on the fulfilment of these requirements. Of course, in practice the extent to which the ALARP principle is implemented is also an issue, but as there are no strict limits to look for, supervision of its implementation is more difficult.

In the following we discuss the ethical justification for such a regulation regime based on the use of minimum requirements, in the form of specific requirements related to arrangements and risk acceptance criteria. The emphasis is on the risk acceptance criteria. Does this regime have a stronger ethical justification than other regimes that do not include risk acceptance criteria as a part of their framework? What conditions need to be fulfilled to obtain such justification? In the Norwegian offshore industry the operators define the risk acceptance criteria. Would that violate the basic idea of minimum requirements, as the operators could specify criteria that in practice are always met?

When discussing the ethical justification for such a regime we have to distinguish between various types of ethics. Two basic directions are (1) ethics of the mind – focusing on the purpose, meaning or intention of the action, and (2) ethics of the consequence – focusing on the good or bad results of an action, see Hovden (1998) and Cherry and Fraedrich (2002). These types of ethics are labelled deontological and teleological theories, respectively. A variant of the teleological theory is utilitarianism, which searches for alternatives with the best balance of good over evil. The use of cost-benefit analysis may be seen as a way of making the theory operational. Deontological theories stress that the rightness of an act is not determined by its consequences. Certain actions are correct in and of themselves because they stem from fundamental obligations and duties (Cherry and Fraedrich 2002).

A regime based on HES requirements and the use of risk acceptance criteria as used by the PSA, is often linked to the former type, the ethics of mind, whereas the use of the ALARP principle is linked to the latter type, the ethics of consequences. The point is that in the former case, the requirements should in principle be fulfilled without reference to other attributes such as costs, whereas in the latter case, decision-making is based on a consideration of all consequences of the possible alternatives.

However, a further look into this way of reasoning shows that it is problematic. When it comes to safety, what are the consequences – the expected outcomes from an activity assigned by some analysts, or the real outcomes generated by the activities? And how do we measure the value of these consequences? For example; how good is a cost reduction compared with a reduction in safety level? This is discussed in more detail below, based on different ethical stands; the duty and the utility stands, but also the justice and discourse stands. The justice approach to ethics focuses on how fairly or unfairly our actions distribute benefits and burdens among the members of a group – people should be treated equally unless there are morally relevant differences between them. The discourse stand is based on a search for consensus through open, informed and democratic debate (Hovden 1998).

In the following discussion we make a sharp distinction between the possible outcomes, uncertainty assessments of what the outcomes will be, and our valuation of the outcomes and quantities expressed through the uncertainty assessments.

2.7.1 The Influence of the Risk Perspectives Adopted

Consider the regulation of an activity, involving a potential for hazardous situations and accidents leading to loss of lives and injuries. From an ethical point of view, we would require no fatalities and no injuries. This is ethics of the mind, no one should be killed or be injured in his or her job. However, in practice no one can guarantee a 100 percentage safety, and alternatives are sought. Examples include the following:

1. The individuals concerned feel safe
2. The individual risk is sufficiently low
3. The calculated individual risk is sufficiently low
4. Risk is reduced to a level that is as low as reasonably practicable
5. The uncertainties related to possible situations and events leading to loss of lives and injuries are reduced to a level that is as low as reasonably practicable.

To discuss these goals and criteria, we need to distinguish between different perspectives on risk, as the meaning of the goals and criteria is different depending on the perspective. Here we restrict attention to two main categories of perspectives:

- a traditional (classical) approach to risk and risk analysis, and
- a Bayesian perspective, see Section 2.1.

Either one starts from the idea that risk (probability) is an objective quantity and this risk has to be estimated, or one starts from the idea that risk (probability) is a subjective measure of uncertainty as seen through the eyes of the analyst. The former case, which is referred to as the traditional or classical view, means that risk is a fictional quantity, expressing the proportion of fatalities in an infinite reference population of similar situations. The latter case is referred to as Bayesian, and has no reference to such an underlying population. Note that there exist many variations of the Bayesian paradigm – here we use the term when probability is used as a subjective measure of uncertainty, see Section 2.1 and the appendix.

A Traditional Approach to Risk and Risk Analysis

We first look at the case when risk acceptance criteria are used.

Traditionally, risk has been seen as an objective property of the activity being studied, and hence there exists an objective real individual risk expressing the probability that the person will be killed or injured. If this probability is low, the person will normally also feel safe. If it can be verified that the real individual risk is below a certain value, the regulator would have ensured that the activity is acceptable from a safety point of view. No guarantee can be given that fatalities or injuries would not occur, but the chance would be small and under control. It is still an argument based on ethics of the mind, as it is grounded on a reflection of what is right and not linked to the possible consequences of the action. A typical value used for individual risk is 0.1%, meaning that there should be a maximum of 0.1% probability that a specified individual will be killed due to an accident occurring during the period of one year. This number is used with no reference to the consequences related to, for example, costs.

The idea that such an objective risk exists is the basis for the regulations in many countries. It is seldom or never explicitly stated, but it is clear from the way the regulations are formulated that such a perspective on risk is adopted.

As an alternative to the above regime based on risk acceptance limits, consider a regulation regime based on the same principles 1–5 above, but with no use of pre-defined risk acceptance criteria. The justification for such a regime would be partly ethics of the mind and partly ethics of the consequences:

- Ethics of the mind: the basic idea, what is a correct risk level for the individual, has to be seen in a broader context taking into account what he or she, and others, gain by the activity. A low accident risk has no value in itself.
- Ethics of the consequences: the specific choice, the action or decision, needs to reflect what the possible consequences are. For example, an alternative may be generated which leads to high risks for some people but extremely positive benefits for others, and the risks can be compensated for through remuneration and insurance.

Of course, even in the case of risk acceptance criteria, the ethical justification is partly teleological: we have to look at the consequences. Requiring a risk level equal to, say, 0.01% would have severe consequences and in many cases mean that activities are not performed. If we adopt the traditional level 0.1%, it is known from many years of experience of using this criterion that it is met for most or nearly all types of activities in the western world.

In addressing a new type of situation, where we have little or no experience from previous studies, it is difficult to specify the risk acceptance criteria. We simply do not know the consequences. An example would be a unique operation, one of great importance. Then using the ethics of the mind to specify a certain limit, would be difficult to justify as the consequences need to be addressed.

The classical approach is based on the idea that an objective risk exists, but in practice we have to estimate this risk, and this estimate would normally be subject to large uncertainties. And this uncertainty needs to be taken into account. Using a risk acceptance criterion of the form 0.1% and adopting a classical view, means that uncertainties in the risk analysis estimate need to be addressed. The true risk number could be significantly different from the estimate. Hence by adopting a regime based on risk acceptance criteria, no minimum requirements have been established, as meeting the 0.1% level does not mean to say that the true risk meets this level. We may try to express the uncertainties of the estimates, but that leads to such complex analysis and such wide uncertainty intervals in most real life cases that the whole idea of using risk analysis and risk acceptance criteria breaks down, see Aven (2003).

An alternative is to refer to Procedure 3 above: specify a limit for the calculated individual risk. However, this would not be satisfactory as there is no guarantee that the real risk is under control.

A Bayesian Perspective

According to the Bayesian view, an objective individual risk does not exist. Using a risk acceptance criterion of the form 0.1% means that the risk analysts' assessment concludes that risk is acceptable or unacceptable, depending on the result of the analysis. However, different assessments could produce different numbers depending on the assumptions made and the analysts chosen for the job.

The idea of minimum requirements defined by the risk acceptance criteria seems to lose its meaning when risk does not exist as an objective quantity. But a further analysis reveals that the Bayesian perspective is not so different from the classical approach, if we acknowledge that in the classical case we have to deal with risk estimates and in the Bayesian case subjective assignments. It is possible to use risk acceptance criteria also in the Bayesian case, interpreting the criteria as limits for comparing the risk assignments. Except for the Criterion 2 stated in Section 3.1, we can implement the others, *i.e.*, 1 and 3–5, with and without pre-defined risk acceptance criteria. The ethical justification would be as in the classical case, interpreting risk according to the Bayesian perspective. We will discuss this in more depth in the following section.

2.7.2 Discussion

It is obvious from the above considerations that the results generated by the risk analysis need to be seen in a broader context taking into account that the risk analysis depends on assumptions made, the analysts performing the analysis *etc.* An ethical principle, based on ethics of the mind, for adopting a pre-defined level, can still be put forward, but the limitations of the analyses weaken its position. We may formulate a risk acceptance limit as a minimum requirement, but the tool to be used to check its fulfilment or not lacks the accuracy or precision needed. To compensate for this lack of accuracy or precision, we could specify a fairly stringent requirement, say 0.01%, and use the analysis to check that this level is fulfilled, as a guarantee for the real risk to be lower than 0.1% (say) in the classical case, or as a guarantee that different analyses would all ensure a level of 0.1% (say) in the Bayesian case. However, such a strong limit would not be used, as the consequences might easily be unacceptable, as discussed in the previous section. Instead a weak limit would be preferred, such as 0.1%, and then the calculated risk would nearly always meet this limit. A minimum safety level is then established, but this level is so weak that it is seldom or never challenged. A lot of energy and resources are used to verify that these limits are met, which is not very cost-efficient as the results are obvious in almost all cases.

The use of conservative assumptions leading to overestimation of risk or higher risk assignments than the "best judgements" made by the analysts is often seen in practice. However, this does not add anything new to the above reasoning, except that such a procedure could simplify the analyses. If the criterion is not met in the first run of the risk analysis, it is necessary to perform further detailing and remove some of the conservative assumptions, which normally leads to acceptance.

At the point of decision-making the consequences X, representing for example the number of fatalities, are unknown. Expectations, $E[X]$, may be calculated and uncertainties assessed, but there is a fundamental difference between the real outcomes and the predictions and uncertainty assessments. The fact that we do not know the outcomes means that we cannot simply apply the ethics of the consequences. In the case of large uncertainties in the phenomena being studied, the weight on the ethics of the mind would necessarily also be large. We can calculate individual death probabilities and expected net present values in a cost-benefit analysis – however, there would be a need to see beyond these calculations, as they are based on a number of assumptions. How to deal with the uncertainties has to have a strong component of ethics of the mind. We are led to the adoption of principles such as the cautionary principle, saying that in the face of uncertainties, caution should be a ruling principle, and the precautionary principle, saying that in the case of lack of scientific certainty about the consequences, the activity should be avoided or measures implemented, see Section 2.3. These principles are primarily principles of ethics of the mind. They are obviously related to the consequences in the sense that they are implemented to avoid negative consequences, but the basic ideas of using these principles are founded in a belief that in the face of uncertainties, caution and precaution should be the ruling policy – you should not gamble with lives.

Adopting a classical perspective on risk, we would add that uncertainties in the risk estimates are another reason for adopting the cautionary and precautionary principle, and emphasise the ethics of the mind. In a way these uncertainties are mind-constructed uncertainties, just as the objective underlying risks are mind-constructed quantities, and in this sense the ethics of the mind may be given too strong a weight compared with the ethics of the consequences.

To evaluate the uncertainties, risk analysis constitutes a key instrument, but also risk perception plays a role. If people perceive the risks and uncertainties related to a phenomenon as high, it could influence the decision-making process and the weighting of the various concerns. However, taking into account risk perception in the decision-making process does not necessarily mean that more emphasis is placed on the ethics of the mind, relative to the ethics of the consequences, as risk perception is also a consequence or an outcome of the actions and measures considered. Depending on the perspectives on risk adopted, risk perception provides to varying degree relevant information about the consequences. If risk is considered an objective property of the system being analysed, risk perception would in general be given less attention than if risk is a subjective measure of uncertainty.

So far we have focused on individual safety. Now some words about environmental issues. Here the ideal would be no damage to the environment. Since this ideal cannot be achieved fully in most cases, the concepts of risk and uncertainty need to be addressed. The first issue we then would like to discuss is whether a low environmental risk level has a value in itself. Clearly a life has a value in itself, and most people would conclude that the environment has a value in itself. But the value is not necessarily very large. If we have to choose between production of a certain type of units causing some risk of pollution, or not, we often accept the risk of pollution. The benefits outweigh the negative consequences. We adopt ethics of the consequences. However, as for lives, we also use ethics of the mind, in the face of risks and uncertainties, as we cannot picture the exact consequences of an action. We apply the cautionary and the precautionary principles. The discussion on whether to use risk acceptance criteria or not would be analogous to the discussion for the individual risk.

In general, the regulators might be expected to put more emphasis on the uncertainties and the ethics of the mind than the industry, as the regulators necessarily have a broader societal perspective. This creates a dilemma. Modern safety management is based on the use of the internal control principle, saying that industry has full responsibility for its activities. In the Norwegian oil industry, this principle has been implemented and the oil companies specify the risk acceptance criteria. However, the primary goal of the industry is profit. In practice, the industry would seek to avoid "unnecessary" constraints in the optimisation process, and hence reduce the ethics-of-mind-based criteria to a minimum. If the regulations require such criteria, the result would be the implementation of very weak limits, so that the criteria do not induce any practical constraints.

Thus the regulators need to specify the criteria if they wish to implement a certain safety standard in the industry. To some extent this is being done in the offshore industry. For example the Norwegian and UK petroleum authorities have defined upper limits for the frequencies of impairment of specific safety functions, see Aven and Vinnem (2005). Unless obliged to do so by the regulator, one should not expect to define risk acceptance criteria beyond these limits, as such criteria might be viewed as being in conflict with the primary goals of the industry. The ethics of the consequences would necessarily rule.

Finally, some words about the justice and discourse stands.

According to the justice principle of ethics, people should be treated equally unless there are morally relevant differences between them. An application of this principle to safety and risk in society and industry is meaningless, as safety/risk is just one of many attributes that define welfare and are relevant in the decision-making process. Specifying some minimum safety standards does not imply full implementation of the principle, but provides some constraints for optimisation. However, how these minimum safety standards should be defined cannot be deduced from an ethical principle. The use of risk acceptance criteria is one way of making such standards operational, but there are other approaches that could be used as well, as discussed above.

The discourse principle is based on a search for consensus through open, informed and democratic debate. Many aspects of this principle are widely implemented in the western world, through modern regulation and management regimes, emphasising involvement and dialogue. Applying this principle is mainly based on the ethics of the mind, as it is believed that this is the right way of dealing with risks and uncertainties.

2.7.3 Conclusions

Many people, in particular safety people, mock the utility principle see Hovden (1998). What they often do is argue against the practical tool for implementing the principle, the cost-benefit analyses. And that is easy. Any tool used for balancing pros and cons would have severe limitations and be based on a number of assumptions. There is therefore a need to see beyond the tools. We need some managerial review and judgement which open up for a broader perspective, reflecting the limitations and assumptions of the tools and all the ethical concerns that need to be taken into account when making decisions in the face of uncertainties. The utility principle (ethics of the consequences) would be important, because we need to balance the pros and cons. However, it is not possible to make this principle operational without also reflecting the other ethical principles (ethics of the mind, justice and discourse). There should be no discussion on this. What can and should be debated is the balance of the various principles and concerns. For example, in the case of helicopter shuttling between offshore installations, what should be an acceptable safety or risk level for the workers? Should we impose some limitations on the number of flights to manage the level of safety for the personnel? Yes, in practice this is done, and it seems ethically correct. The argument is of course based on considerations of the consequences, but also ethics of the mind and the justice principle – the workers should be ensured a certain safety level. As a result of the regulation and management regime, processes have been implemented involving the workers in specifying this level. To the greatest possible extent, consensus is sought.

Observe that the specification of such a safety level (for example expressed by a maximum number of flights n) is not the same as using a pre-determined risk acceptance criterion, as discussed above. The decision to be taken is to determine n and for that purpose a procedure emphasising the generation of alternatives (different n values) and assessment of the associated consequences, may be adopted. This process generates a specific solution, the proper n. The shuttling risk analysis is used as a basis for specifying this n.

Of course, one may decide that the associated risk should be reformulated as a risk acceptance criterion to be used for other applications. If that is the case, the discussion of the previous two sections applies. The ethics of the mind is highlighted relative to the ethics of the consequences. However, as discussed in the previous two sections it is possible to also formulate procedures according to the ALARP principle that are strongly motivated by the ethics of the mind – the point is that significant uncertainties in the consequences cannot be adequately handled by standard cost-benefit analyses. Applications of the cautionary and precautionary principles are required.

As discussed in this section, we see no stronger ethical arguments for using predefined risk acceptance criteria in preference to any other regimes. There are obviously arguments for using and not using any of the above regimes, but these are not primarily of ethical character. To decide which regime to implement, ethical considerations should obviously be taken into account, but the decision has to be put into a wider context reflecting the practical implementation of the regimes and how to understand and deal with risk and uncertainty. Most people would agree that the chosen regime must balance a number of concerns and ethical perspectives. The aim of this discussion has been to contribute to clarification of this context and provide an improved basis for performing this balance.

Bibliographic Notes

Section 2.1 and 2.2 are partly based on Aven (2003), and Abrahamsen *et al.* (2005), Abrahamsen and Aven (2006a) and Aven and Kristensen (2005). There are many examples presented in the literature showing that people violate the axioms of the expected utility theory. A key reference is the Allais paradox, see French and Insua (2000). Several modifications and extensions of the expected utility theory have been suggested, to better reflect people's actual choices and preferences. One of the most popular is the rank dependent utility theory, see Yaari (1987) and Wakker (1994). However, we have a firm and societal perspective, and then the axioms of the expected utility theory make sense in most cases.

Section 2.3 is based on Aven (2006a) and Aven and Abrahamsen (2006). The precautionary principle has been thoroughly discussed in the literature. In addition to the cited references in Section 2.3, we refer to Rodgers (2001), Morris (2000), Pearce (1994), Wiener and Rogers (2002), EU (2001), Gollier *et al.* (2000), Gray (1990) and HSE (2001a).

Section 2.4 is a summary of Abrahamsen *et al.* (2004).

Section 2.5 is based on Sandøy *et al.* (2005), which in turn is based on Abrahamsen *et al.* (2004), Aven *et al.* (2004) and Aven and Kristensen (2005).

Section 2.6 is based on Aven and Vinnem (2005) and Aven *et al.* (2006b, c). For a review of how risk acceptance criteria and the ALARP principle have been used in the Norwegian and UK offshore oil and gas industries, we refer to Hokstad *et al.* (2003), Aven and Pitblado (1998), Norsok (1999), Vinnem (2000) and Vinnem *et al.* (1996, 2006c). The ALARP principle is discussed in these references as well as in UKOOA (1999), Melchers (2001), Pape (1997) and Schoefield (1998). For additional references covering risk acceptance and risk acceptance criteria we refer to Aven (2003), Fischhoff *et al.* (1981), Lind (2002), Rimington *et al.* (2003), Skjong and Ronold (2002), The Royal Society (1992) and Vatn (1998).

Evans and Verlander (1997) and Abrahamsen and Aven (2006b) provide examples showing that the use of risk acceptance criteria can lead to unreasonable conclusions and inconsistency in decision-making. Abrahamsen and Aven (2006b) uses the axioms of the expected utility theory and the rank dependent theory as the reference.

Figure 2.1 could be associated with the specific procedure for using cost-benefit analyses or decision analyses as discussed in Fischhoff *et al.* (1981), *cf.* also Hertz and Thomas (1983). Our approach is however based on a more pragmatic view on the use of formal analyses, reflecting real-life decision processes.

Ethical justification for safety investment has been thoroughly discussed in the literature. In addition to the Hovden reference cited above, we refer to HSE (2001a), Hattis and Minkowitz (1996), Hokstad and Steiro (2005) and Shrader-Frechette (1991).

Section 2.7 is based on Aven (2006b).

3

A Risk Management Framework for Decision Support under Uncertainty

3.1 Introduction

The previous chapters have shown the need for providing guidance and a structure for decision-making in situations involving high risks and large uncertainties. To this end we present in this chapter a framework for risk management and decision-making under uncertainty. The framework comprises the following main elements

- problem definition (challenges, goals and alternatives),
- stakeholders,
- concerns that affect the consequence analyses and the value judgements related to these consequences and analyses (frame conditions and constraints),
- identification of which consequence analyses to execute and execution of these,
- managerial review and judgement,
- and the decision.

Risk is defined as the combination of the two basic dimensions: a) possible consequences and b) associated uncertainties. As there are many facets of these dimensions, the framework means a broad perspective on risk, reflecting for example that there might be different assessments of uncertainties, as well as different views on how these uncertainties should be dealt with. This is also reflected by the risk classification scheme adopted, based on the system introduced by Klinke and Renn (2000) and modified by Kristensen and Aven (2004).

The framework gives a structure for classification of HES decision problems, and a procedure for execution of the related decision-making processes. It provides a check list for what concerns to address when searching for the best decision alternative. The classification is based on the two dimensions expected consequences and uncertainties.

The aim is to obtain better decisions *i.e.*, to obtain more desirable outcomes, but of course this is hard to evaluate. The best we can do, at least in the short run, is to increase confidence in being able to obtain desirable outcomes. Furthermore, our work would optimise the time and cost required involved in the process.

The chapter is organised as follows. In Section 3.2 we present the decision framework, after having summarised some fundamental principles of the framework in Section 3.1. A discussion of the framework is found in Section 3.3, together with some conclusions.

3.2 Basic Building Blocks of the Framework

The framework is based on a set of building blocks as summarised below. These building blocks are extracted from the review and discussion of the fundamental issues in Chapter 2.

(a) Risk is in general characterised by the combination of possible consequences associated with an activity and the assessor's uncertainty about these consequences. The consequences are normally expressed by quantities that can be measured (such as money, loss of lives, *etc.*). A set of quantities are typically needed to give a proper description of the consequences. We refer to these quantities as observable quantities or just observables.

If C represents the consequence and c describes one possible value of C (or an interval defined by c, for example $[0,c]$), risk is expressed by the combination of possible c values and our uncertainty as to the consequence C will take the value c.

(b) Risk (uncertainty) is quantitatively expressed by probabilities and expected values. We assess the uncertainties and assign probabilities (and hence we assign values for risk). It is meaningless to speak about uncertainties in assigned probabilities and risk numbers, as these values express uncertainties, conditional on some information and knowledge.

(c) Risk analyses provide decision support, by analysing and describing risk (uncertainty). The risk analysts analyse the risks, and evaluate them, *i.e.*, they discuss the significance of the risks, in relation to comparable activities and possible criteria. The combination of risk analysis and risk evaluation is referred to as a risk assessment. The analyses need to be evaluated in light of their premises, assumptions and limitations. The analyses are based on background information that must be reviewed, together with the results of the analyses. The decision-maker performs what we refer to as a managerial review and judgement.

(d) A sharp distinction is made between facts, risk assignments, risk evaluation, and risk treatment, where risk treatment means the process of selection and implementation of measures to modify risk.

(e) It is essential to make a sharp distinction between what are expected values determined at the point of decision-making and what the real observations (outcomes) are. The expected values give, to varying degree, good predictions of the future observations. Uncertainty and safety management are justified by reference to these observations and not the expected values alone.

(f) Proper uncertainty management and safety management seek to produce more desirable outcomes, by providing insights about the uncertainties relating to the future possible consequences of a decision, and controlling and reducing these uncertainties.

(g) A decision rule based on the expected *NPV* with a risk-adjusted discount rate or risk-adjusted cash-flows, should be supplemented with uncertainty assessments to see the potential for uncertainty and safety management in later phases.

(h) What is acceptable risk and the need for risk reduction cannot be determined just by reference to the results of risk analyses. To be precise, we do not accept a risk but a solution, with all its attributes.

(i) Cost-benefit analysis means calculating expected net present values with a risk-adjusted discount rate or risk-adjusted cash-flows. In a societal context, society's willingness to pay is the appropriate reference, whereas for businesses, it is the decision-maker's willingness to pay that is to be used.

(j) Cost-effectiveness analysis means calculating measures such as the expected cost per number of expected saved lives.

(k) A multi-attribute analysis is an analysis of the various attributes (costs, safety, ...) of the decision problem, separately for each attribute.

(l) Risk and decision analyses need extensive use of sensitivity and robust analyses.

Thus we adopt a broad perspective on risk, acknowledging that risk cannot be distinguished from the context it is a part of, the aspects that are addressed, those who assess the risk, the methods and tools used, *etc*.

We define the term vulnerability as the combination of possible consequences and associated uncertainties given a source, *i.e.* given a threat, hazard or opportunity. These three source categories are typically used in security, safety and economic contexts, respectively. Security relates to intentional situations and events. An example of an 'opportunity' is a planned shutdown, which allows for preventive maintenance.

Based on this definition, we refer to 'a vulnerability' as an aspect or feature of the system, when the combination of possible consequences and associated uncertainties is judged to give a high vulnerability, *i.e.*, is considered critical in some sense. For example, in a system without redundancy the failure of one unit may result in system failure, and consequently we may judge the lack of redundancy as a vulnerability depending on the uncertainties.

The issues (e) to (g) are related to the manageability of the risk. Some risks are more manageable than others, meaning that the potential for reducing the risk is larger for some risks than for others. The concept is illustrated in Figure 3.1.

Alternative 1 gives a medium risk level and low manageability, whereas alternative 2 gives a higher risk but also a higher manageability. Thus by selecting alternative 2 a higher risk is initially assigned, but it provides a large opportunity for reducing the risk and obtaining good safety performance (by adopting a good safety management).

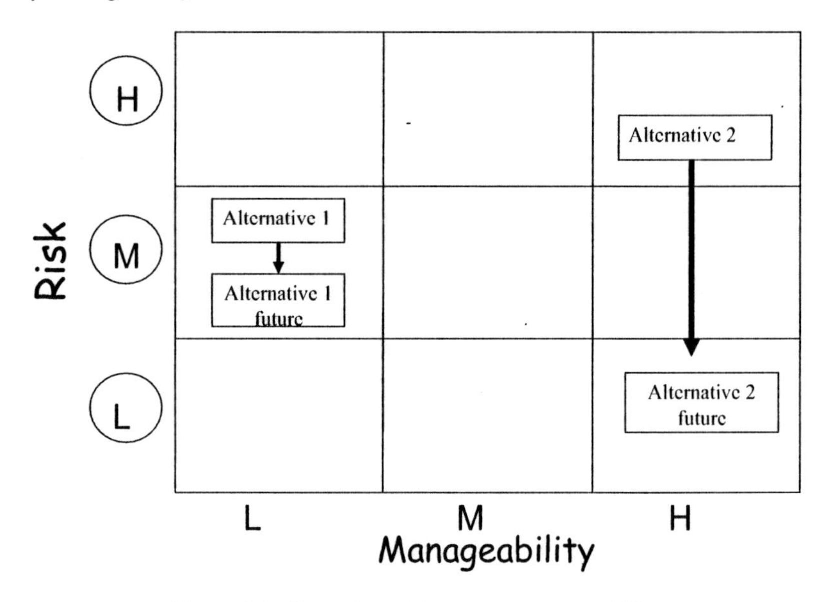

Figure 3.1. Illustration of the concept manageability

Figure 3.2 illustrates how the degree of uncertainties and manageability depends on the phase of development. If our focus is observables $X = (X_1, X_2, \ldots)$, we predict X, normally using the expected value $E[X|K]$, where K is the background information (knowledge). The degree of uncertainties and manageability is large at an early stage of development, and decreases as a function of time. Of course, this is a schematic illustration, showing typical trends in practice, we may have situations where for example the uncertainties increase. Because of large uncertainties the outcomes of X may deviate strongly from the predictions. However, by proper uncertainty management, and safety management the goal is to obtain desirable outcomes.

Following our definition of risk, a low degree of uncertainty does not necessarily mean a low risk, and a high degree of uncertainty does not necessarily mean a high level of risk. This is important. As risk is defined as the combination of possible consequences and the associated uncertainties (quantified by probabilities), any judgement about the level of risk, needs to consider both dimensions. For example, consider a case where only two outcomes are possible, 0 and 1, corresponding to 0 and 1 fatality, and the decision alternatives are A and B, having uncertainty (probability) distributions (0.5,0.5), and (0.0, 1.0), respectively. Hence for alternative A there is a higher degree of uncertainty than for alternative B. However, considering both dimensions, we would of course judge alternative B to have the highest risk as the negative outcome 1 is certain to occur.

The above building blocks constitute a basis for the framework presented. They are premises for the work and their justification and suitability will not be discussed in this chapter.

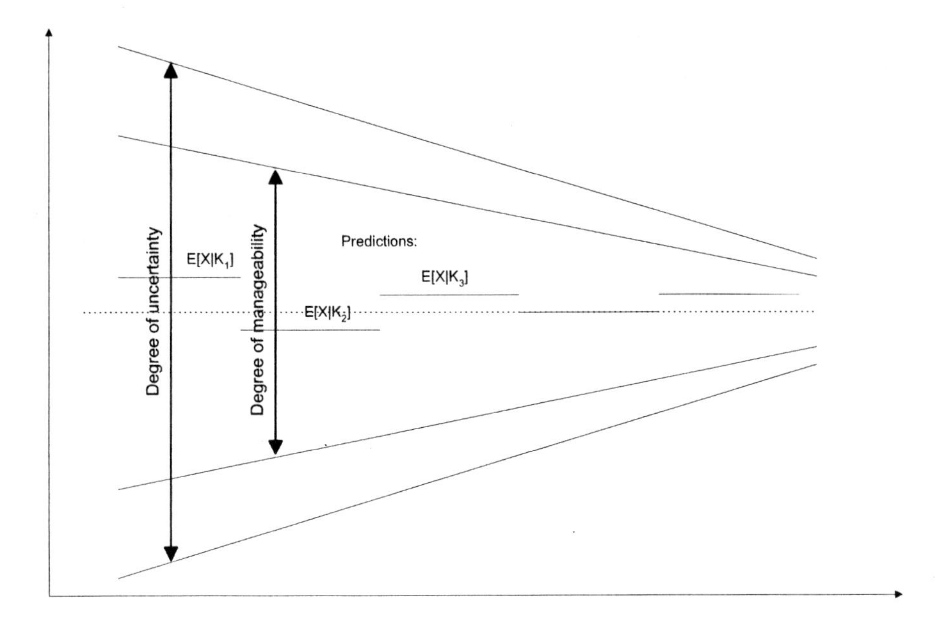

Figure 3.2. Illustration of the level of uncertainties and manageability as a function of time

3.3 The Framework

Our starting point is a decision-maker facing some decision points in a project. These decision points include problems and opportunities, such as poor HES results, implementation of a risk reduction policy, the use of new technology, choosing a concept for further evaluations, *etc*. Having identified the main decision points, adequate decision alternatives need to be generated and evaluated, relating to whether or not to execute an activity, alternative concepts, design configurations, risk reducing measures, *etc*. Our focus is on situations characterised by a potential of rather large consequences, large associated uncertainties and/or high probabilities of what the consequences will be if the alternatives are in fact realised, *i.e.*, high risks according to our definition of risk. The consequences and associated uncertainties relate to economic performance, possible accidents leading to loss of lives and/or environmental damage, *etc*. Risk and decision analyses are considered to give valuable decision support in such situations.

The framework proposed for risk-informed decision-making and HES management is based on the idea that HES considerations should be included actively in the decision process and not only be viewed as a frame condition for other business activities. This means that HES aspects and attributes related to HES must be included in the key decision-making processes that determine the choice of concept, the design configurations *etc*. HES considerations have to be included in the broader context of risk and uncertainty management, also covering economic analysis and opportunities.

The proposed framework recognises the fact that individual decision problems may differ considerably in relation to potential consequences and associated uncertainties of what will be the consequences, *i.e.*, the risks. The differences in management and decision level may also be large, ranging from corporate level to government and international organisations such as UN. The result is a need for different types of decision support and decision-making processes. The framework introduces a classification system structuring the different situations, to select among alternative methods for, and approaches to, risk-informed decision-making, see Section 3.3.2.

Figure 3.3 illustrates the principles of the proposed framework for HES management.

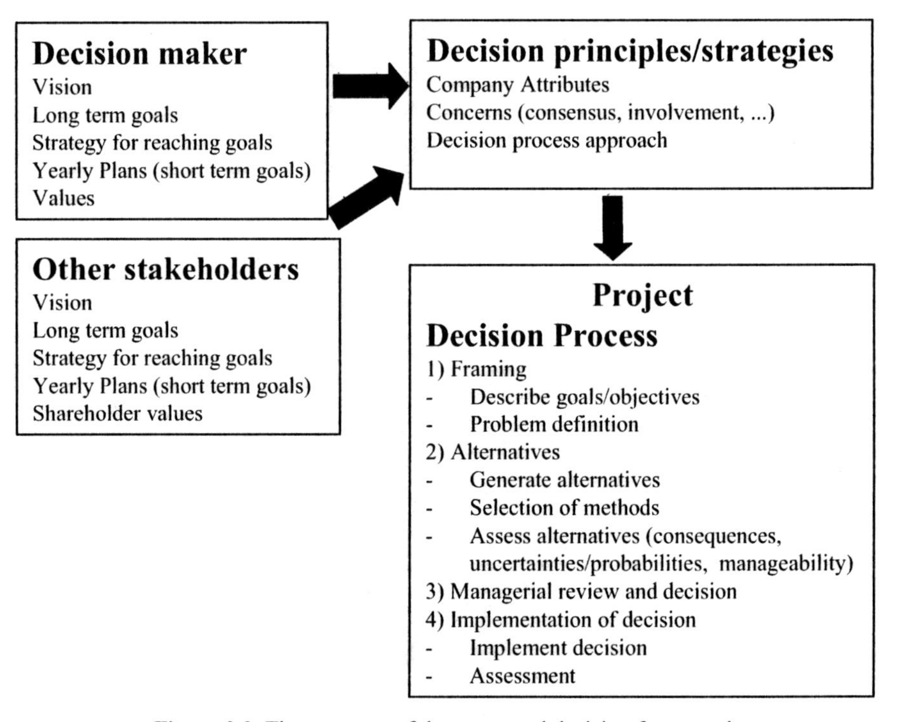

Figure 3.3. The structure of the suggested decision framework

The figure shows that the decision-maker and the other stakeholders influence the decision principles and strategies, and further that these principles and strategies affect the decision process. For each project there may be several decision problems requiring a structured approach to decision-making, including generation and evaluation of alternatives, managerial review and decision-making. The following sections describe the framework and its elements in more detail.

3.3.1 Decision-maker and Stakeholders

The decision-maker could be a person, a board of directors in a firm, the government, *etc.*, making the decision. A stakeholder is any individual, group or organisation that may affect or be affected by, or perceive itself being affected by the decision. Internal stakeholders could be the owner of an installation, other shareholders, the safety manager, unions, the maintenance manager, *etc.*, whereas external stakeholders could be the safety authorities (the Petroleum Safety Authority, the State Pollution Control Agency, *etc.)*, environmental groups (Green Peace, *etc.)*, research institutions, *etc.*

The stakeholders, including the decision-maker, have defined visions and long-term goals for their activities, and strategies and plans for meeting these visions and goals, reflecting important concerns and values for the stakeholders.

3.3.2 Decision Principles and Strategies

- The values, visions and goals, strategies and plans of the decision-maker and other stakeholders are the basis for forming the high-level decision principles and strategies, to steer the decisions in the desired direction. Examples of such principles and strategies are: a defined list of attributes to be evaluated, for example, economy, HES, societal responsibility, reputation *etc.*
- High-level principles of decision-making, such as: a search for consensus, third party involvement, the use of the cautionary and precautionary principles, *etc.*
- An overall procedure for how to perform the decision-making processes
- A procedure for implementing the ALARP principle for risk to personnel, environment and assets, *etc.*

These principles are adopted for all projects and decisions within the company or organisation concerned.

3.3.3 Decision-making Process

This section presents the stepwise decision-making process shown in Figure 3.3. The focus in this will be on step 2; alternatives. The starting point of the decision process is a crude description of the project or case being studied.

Framing
Description of Goals and Objectives. The definition of goals and objectives is a key element of the decision-making process. At this stage in the process we are concerned about high-level goals, relating the project activities and the decisions to the goals and objectives of the company or the organisation that the project is a part of.

Definition of the Decision Problem. In this activity focus is on the specific decision problem considered. What is the problem to be addressed and solved, and what are the results to be obtained by making the right decision? Furthermore, what are the frame conditions, including relevant criteria and requirements for solving the problem? These criteria and requirements may relate to production, HES, functionality, quality *etc.*

Generation and Evaluation of Alternatives

Generation of Alternatives. This activity means generating a list of alternative solutions to the described decision problem. Sometimes this is quite a simple process, as there are some obvious candidates, for example if the problem concerns whether to implement a measure or not. In other cases there could be many possible solutions to the defined problem and each solution may have a subset of alternatives that need to be evaluated.

The process of generating alternatives is typically arranged as a creative workshop with relevant personnel attending, leading to a list of alternatives for further evaluation.

Selection of Method for Risk-Informed Decision Assessment. The method used for evaluating alternatives will depend on the type of decision problem in question. Some decisions are critical and need detailed analysis, whereas others are not so critical and more crude analyses may be sufficient. To adapt to this variation, we have designed a characterisation method to select among the alternative methods for decision-making.

The purpose of introducing this characterisation process is to be able to select the most efficient decision process for the relevant decision problem, using a structured and standardised process that ensures the quality and the documentation of the decision process. The following three main categories of decision problems are introduced:

(1) Standard decision problem:
- This category applies for most decisions.
- The decision problem is characterised by limited expected loss/gain and limited uncertainties.
- The project management team facilitates the decision process without external assistance; there is no need for detailed quantitative analysis.

(2) Advanced decision problem:
- This decision problem is characterised by significant expected loss/gain and significant uncertainties.
- The project management team facilitates the decision process, but there is a need for detailed analysis on selected issues and external expertise will be required.

(3) Complex decision problem:
- This decision problem is characterised by large expected loss/gain and large uncertainties.
- The project management team engages experienced people for facilitation and documentation of the decision process.
- Detailed analysis will be required.

A set of features linked to expected consequences and uncertainties is established as a basis for performing the categorisation, see Table 3.1. The ranking categories obviously need to be tailor-made to the specific application. The system works as follows. For a specific decision alternative, we assess the investment costs, and assign an expected value giving a ranking from 1 to 3 depending on the result. Similarly, we assess the potential profit by implementing the decision alternative. Expected values provide the starting point for assessing the importance of the problem; however, there may be a number of factors that can cause or lead to outcomes (consequences) that are extreme compared to the expected values, and such factors are reflected by the second dimension, uncertainties.

Table 3.1. Characterisation of expected consequences and uncertainties. An illustrative example

	Ranking			Argument	Comment
	1	2	3		
Expected effect of implementing decision alternative					The highest ranking on one of the questions gives the total ranking, ie. a ranking of 3 in one of the questions gives a total ranking of 3 wrt potential.
Investment	>10 mill	10< x < 100	> 100 mill		
Profit/production increase	>10 mill	10< x < 100	> 100 mill		
Production loss	>10 mill	10< x < 100	> 100 mill		
Project schedule	Little	Medium	Large		
Safety (loss of lives)	Little	Medium	Large		
Working environment	Little	Medium	Large		
Environment	Little	Medium	Large		
Reputation	Little	Medium	Large		
Total					
Uncertainties					The highets ranking on one of the questions gives the total ranking, ie. a ranking of 3 in one of the questions gives a total ranking of 3 wrt uncertainty.
Complexity of technology	Low	Medium	High		
Complexity of organisation	Low	Medium	High		
Availability of information	High	Medium	Low		
Time frame	Short	Medium	Long		
Vulnerabilities	Low	Medium	High		
Total					

The following rules are proposed for the total ranking of the two dimensions:

- **Expected consequences:** The highest ranking for one of the issues gives the total ranking *i.e.*, a ranking of 3 in one of the questions gives a total ranking of 3, *etc.*.
- **Uncertainties:** The highest ranking for one of the questions gives the total ranking, *i.e.*, a ranking of 3 in one of the questions gives a total ranking of 3.

In theory we should run through this process for each alternative, and the highest scores determine the overall score for the problem. However, in practice we normally conduct only one run integrating the alternatives, and using a typical or the most extreme alternative as a basis for the categorisation.

The above characterisations provide a basis for determining an appropriate classification of the problem. High scores on each dimension mean that the problem is classified as complex, whereas low scores result in a standard decision problem. Of course, other factors than this risk assessment may also affect the final selection, such as time constraints and available resources.

Evaluation of Alternatives

In the following we describe the analysis approach for the different evaluation approaches:

- Standard decision problem
- Advanced decision problem
- Complex decision problem.

Standard Decision Problem. The decision alternatives are evaluated on the basis of checklists, codes and standards. A solution meeting the standards and codes is chosen. In some cases there is a need for ranking alternatives, and then a light version of the advanced decision problem analysis approach may be used.

Advanced Decision Problem. We refer to the complex decision problem. The difference between the two analysis levels is related to the degree of quantification. All the aspects that are covered by the complex decision problem are also covered by the advanced decision problem. However, for the advanced decision problem, more crude analyses are performed providing results in categories, expressing different levels of risks, costs, *etc.* A comparison of options is performed summarising the pros and cons of the various alternatives, using some type of matrix showing the degree of performance for each alternative and each attribute.

Complex Decision Problem. For this type of analysis a complete quantitative analysis is required. That means that for all relevant attributes, observables are identified. As before we denote these $X = (X_1, X_2, \ldots)$. These may represent costs, incomes, NPV, production volumes, number of fatalities due to accidents, *etc.* By different types of risk and uncertainty analyses, these observables are predicted and uncertainties assessed. Expected values are assigned, but the analyses see beyond the expected values, covering:

1. Specific features of the possible consequences. It is not obvious what quantities X_i should be addressed. Some aspects of the possible consequences need special attention.
2. Specific features of the uncertainties. There may be large uncertainties in the underlying phenomena, and experts may have different views on critical aspects and risks.
3. The level of manageability during project execution. To what extent is it possible to control and reduce the uncertainties, and obtain desired outcomes?

For Category 1 we suggest using the structure by Renn and Klinke (2002) and modified by Kristensen and Aven (2004). The scheme consists of eight consequence characteristics:

(a) Potential consequences (outcomes) – represented by representative performance measures (future observable quantities) such as costs, income, production volumes, deliveries, number of fatalities *etc*.
(b) Ubiquity – which describes the geographical dispersion of potential damage.
(c) Persistency – which describes the temporal extension of the potential damage.
(d) Delay effect – which describes the time of latency between the initial event and the actual impact of damage. The time of latency could be of physical, chemical or biological nature.
(e) Reversibility – which describes the possibility of restoring the situation to the state before damage occurred.
(f) Violation of equity – which describes the discrepancy between those who enjoy the benefits and those who bear the risk.
(g) Potential of mobilisation – which is to be understood as violation of individual, social and cultural interests and values generating social conflicts and psychological reactions by individuals and groups who feel afflicted by the risk consequences. Mobilisation potential could also differ as a result of perceived inequities in the distribution of risk and benefits.
(h) The difficulty of establishing appropriate (representative) performance measures (observable quantities on a high system level).

For Category 2 various types of uncertainty analyses can be used. The risk analysis is to be seen as an uncertainty analysis of future observable quantities and events. The analysis structures the analysts' knowledge of the risks and vulnerabilities and of what the consequences of a hazard could be. Normally, a number of scenarios could develop from a specific hazard. There are uncertainties present, and these uncertainties need to be assessed and described. To assess the uncertainties about the possible consequences we may adopt a classification system as follows, in addition to using probabilities to express the uncertainties related to the outcomes of the various observables:

(a) Insight into phenomena and systems – which describes the current knowledge and understanding about the underlying phenomena and the systems being studied.
(b) Complexity of technology – which describes the level of complexity of the technology being used, reflecting for example that new technology will be utilised.
(c) The ability to describe system performance based on its components.
(d) The level of predictability, from changes in input to changes in output.
(e) Experts' competence – which describes the level of competence of the experts used in relation to for example the best available knowledge.

(f) Experience data – which describes the quality of the data being used in the analysis.
(g) Time frame – which describes the time frame of the project and how it influences the uncertainties.
(h) Vulnerability of system – which describes the vulnerability of the system to, for example, weather conditions and human error. A more robust technical system is more likely to withstand strains and this can reduce the likelihood of negative consequences.
(i) Flexibility – which describes the flexibility of the project and how this affects uncertainty, *e.g.*, a high degree of flexibility allows adjustments to the project plan as more information becomes available and this can reduce the potential for negative outcomes.
(j) Level of detail – which describes the need for more detailed analysis to reduce uncertainty about the potential consequences.

The uncertainty aspects (a)–(j) can be assessed qualitatively, and discussed, or assessed by some type of classification and scoring system, describing the analysts' and the experts' knowledge and judgements.

For Category 3, the task is to assess the potential for obtaining desirable consequences, by proper uncertainty and safety management. By desirable consequences we mean desirable outcomes of the performance measures X, including desirable consequences of features of the consequences such as ubiquity, persistency, delay effect, reversibility *etc*. Such assessments can be qualitative and address factors that are important for obtaining such outcomes. The scientific disciplines would in many cases indicate areas that are important; unfortunately, though, in many cases establishing the appropriate measures is not straightforward. The reason for this is often a lack of knowledge about the effect of the measures, as well as a dispute among experts on the effects of these measures. An example is the importance of obtaining a good safety culture in a company – we all acknowledge that safety culture is important, but the effect is difficult to measure.

The costs of the uncertainty and safety management also need to be addressed and this leads to the use of some type of cost-benefit or cost-effectiveness analyses, measuring, for example, the expected cost per expected number of saved lives. Such analyses provide, in many cases, essential decision support, but additional assessments are often needed, for example, addressing manageability characteristics such as:

(a) The ability to run processes reducing risks (uncertainties) to a level that is as low as reasonably practicable (ALARP),
(b) The ability to deal with human and organisational factors and obtain a good HES culture.

These aspects can be assessed qualitatively, and discussed, or assessed by some type of categorisations and scoring system, describing the analysts' and the experts' knowledge and judgements.

The above assessments provide decision support by structuring and communicating the analysts' knowledge to the decision-maker. As part of this support we may use cost-benefit analyses and other types of analysis, that explicitly reflect the weight the decision-maker gives to the various attributes. However, such analyses still provide decision support – not decisions. There is a need for managerial review and judgement.

Managerial Review and Decision

The extent to which the analyses and evaluations provide clear recommendations on which decision alternative to choose would be case dependent. Purely mechanistic procedures for transforming the results of the analyses and evaluations into a decision cannot be justified. The analyses need to be evaluated in the light of the premises, assumptions and limitations of these analyses. The analyses are based on background information that must be reviewed together with the results of the analyses. The analyses and evaluations provide decision support – not a decision. When evaluating the decision support the decision-maker needs to consider a number of issues, including

- Is the decision-making process managed and documented according to the decision principles and strategies?
- What is the ranking of the alternatives based on the analyses and evaluations? What assumptions are the analyses, evaluations, and ranking based on? What are the limitations of the analyses and evaluations?
- Are there concerns not taken into account in these analyses and evaluations?
- Are all relevant stakeholders taken into account? Would different weights of some stakeholders affect the conclusion?
- Robustness in the decision. What is required to change the decision?

It is often a requirement that the decision support should be available and the decision traceable, which means the documentation of which elements have been given weight in the decision. Such a requirement does not apply in general, the point being that the process is in line with the decision principles and strategies defined. A company may not consider it desirable to trace all weights given to the various attributes.

Implementation of Decisions and Evaluation

How a decision alternative is to be implemented must be considered when evaluating the different decision alternatives.

The effect of a decision alternative could change in time, the response of a measure could be different from that expected, concerns may change, *etc*. The decisions therefore have to be evaluated with hindsight, to see how they performed relative to the challenges and problems they were supposed to meet. From this, experience is gained and modifications in strategies, decision principles, *etc*. may be formulated.

3.4 Discussion and Conclusions

Our classification of decision problems into three categories (complex, advanced, standard; refer to Figure 3.3) may be compared with the classification proposed by UKOOA (1999), consisting also of three categories:

Type A	Nothing new or unusual Well understood risks Established practice No major stakeholder implications
Type B	Lifecycle implications Some risk tradeoffs/transfers Some uncertainty of deviation from standard or best practice Significant economic implications
Type C	Very novel or challenging Strong stakeholder views or risk transfer Large uncertainties Perceived lowering of safety standards

Type A corresponds to standard, Type B to advanced and Type C to complex decision problems. The two classification systems are to a large extent overlapping. Our approach is considered to be more complete and covers a broader range of issues, for example in relation to uncertainty and manageability.

Our Category 3 (complex decisions) may be compared to the approach chosen by Kastenberg *et al.* (2004), which has split the category into two subcategories, in the sense that they consider complicated versus complex systems. According to Kastenberg *et al.*, complicated systems: (1) are understandable by studying the behaviour of their component parts, (2) can be deduced on the basis of cause and effect, and (3) can be determined independent of the observer, that is, deduced only from "objective" empirical observations.

Complex systems, on the other hand, will have to satisfy at least one of the following. They are: (1) holistic/emergent – the system has properties that are exhibited only by the whole and hence cannot be described in terms of its parts, (2) chaotic – small changes in input often lead to large changes in output and/or there may be many possible outputs for a given input, and (3) subjective – some aspects of the system may only be described subjectively. Hence, there may be system properties not exhibited by the parts alone, there may not be a causal relationship between input and output or the output may be path dependent, and no analytic description for the system may be possible.

We have chosen not to adopt this distinction, as other attributes are judged equally important. Our main classification categories are features of consequences, uncertainties and manageability. However, two of the key aspects reflected by the Kastenberg *et al.* distinction are included in the uncertainty factors (c)–(d) for complex decision problems, see above.

Bibliographic Notes

There is a huge body of literature on decision-making and related frameworks. We refer to Watson and Buede (1987), Clemen (1996) and Lindley (1985). For some specific decision frameworks dealing with HES, reference is given to HES (2001), Aven (2003), UKOOA (1999), Hokstad *et al.* (2003), Kastenberg *et al.* (2004), Aven and Korte (2003), Vinnem *et al.* (1996). We also would like to draw attention to the HES Management Regulations issued by the Petroleum Safety Authority (2001).

The chapter is to a large extent based on Aven *et al.* (2006d).

4

Applications – Concept Optimisation

This chapter and the subsequent three chapters present applications of the risk management framework presented in Chapter 3. All applications are taken from the offshore petroleum industry. Some of the applications are taken from decision-making in actual projects, whereas others are inspired by actual projects, but idealised or simplified in order to serve as useful illustrations. Where actual projects have been used, the presentation is made in an anonymous manner, with a few exceptions where all the information is in the public domain.

The text in these four chapters takes a life cycle approach, and there are examples presented from each of the main life cycle phases. The present chapter is focused on the early concept optimisation, *i.e.*, prior to the design phases.

4.1 Historical Background

4.1.1 Ocean Ranger

The semi-submersible mobile drilling unit Ocean Ranger capsized on 15 February 1982 in Canadian waters. The ballast control room in one of the columns had a window broken by wave impact in a severe storm. Short circuits occurred in the ballast valve control systems, when the seawater entered the room, thereby starting spurious operations of the ballast valves. The crew then had to revert to manual control, but was probably not well trained in this and actually left the valves in the open position for some time, when it had been assumed that they were in the closed position. Correction of this failure did not occur sufficiently soon to avoid an excessive heel angle. Due to this excessive heel angle, the rig could not be brought back to a safe state, because only one ballast pump room was provided in each pontoon, at one end. The heel angle was such that the suction height soon exceeded the maximum of 10 metres, and water from the lowest tanks could not be removed.

The onshore based SAR helicopters could not assist due to the severe weather conditions involving strong wind and low visibility. The rig therefore capsized and sank before any assistance could be provided.

Only one lifeboat with less than half of the crew was ever sighted. But both lifeboats appeared to have been launched, and it had to be assumed that all of the personnel (84 man crew) at least attempted to evacuate. The fate of the second lifeboat was never discovered. The lifeboat which was sighted collided with the standby vessel during the transfer attempt from the lifeboat onto the deck of the larger vessel. Within a short time the boat started to drift away, and was never seen again. No survivors or bodies were ever found.

There are some very important lessons to be learned from this accident, not least regarding operational safety. The main lessons are the following:

- Ballast pumping needs system flexibility in order to enable rectification of serious accidental conditions in unforeseen circumstances.
- Competence and training are important in order to enable manual control when automatic systems fail.
- Conventional lifeboats (whether one or two was used is unknown) could apparently be launched even in bad weather conditions.
- Rescue of people from lifeboats by traditional vessels was virtually impossible in bad weather conditions, without special equipment.

There were also some near-misses that occurred about the same time in the North Sea, which underlined the potential seriousness of this issue. In fact, the capsize and sinking of the flotel "Alexander L. Kielland" in 1980 after a severe structural failure had also underlined the need for an extra barrier. More details about these accidents can be found in Vinnem (2007).

The Norwegian Maritime Directorate (NMD) changed their regulations from the early 1980s, as a consequence of this and some other accidents. A requirement was included for reserve buoyancy in the deck of semi-submersible units. This buoyancy is designed to be an extra barrier against capsizing, if extensive water filling of several ballast compartments should occur.

4.1.2 Legislative Situation

It is usually the case in Norwegian offshore regulations, that technical requirements are not given retrospective application, so that installations designed according to an older set of regulations will not need to be upgraded when stricter requirements are stipulated. In this case, however, retrospective application was implemented, so that existing units also had to comply with the new requirements after a transition period. Existing units mainly had to fulfil the requirements by installation of external buoyancy tanks, because it would be impossible to retrofit watertight compartments in the deck structure.

Mobile offshore units, including floating production units, have, for more than 20 years (since the early 1980s), been designed with reserve buoyancy in the deck structure, according to NMD regulations. New mobile offshore drilling units appear to be designed according to NMD regulations (in addition to other regulations and IMO regulations) as the rule, apparently because being able to operate in Norwegian waters is important.

When it comes to floating production units of the semi-submersible type, there has for some years been a tendency to question the requirement for reserve buoyancy. It may be asked if it is really necessary to install this barrier, or can it be deviated from based upon performance of risk assessments? A few installations have recently been installed without this barrier.

In July 2005 however, the BP operated floating production unit "Thunder Horse" in US Gulf of Mexico experienced a severe listing of some 20°, resulting from a hurricane, in which case the installation had been demanned completely as the precaution is in these waters. Production had not started at the time, and no risers had been connected. When the hurricane moved away, the unit was discovered to have this severe condition, and rapid salvage (deballasting) operations had to start. These were successful and the unit was brought back to a stable condition after a few days. One of the statements made by the operator, however, was that the reserve buoyancy in the deck probably prevented the full capsizing of the installation. This demonstrates clearly how important this extra barrier may be.

4.2 Typical Current Decision-making

The typical approach to decision-making on whether to install reserve buoyancy in the deck or not, is the following: a narrowly-restricted evaluation of risk to personnel in relation to risk acceptance criteria, without any further considerations. The quantitative results could be as follows:

- Average FAR value without reserve buoyancy: 5.9
- Average FAR value with reserve buoyancy: 5.3
- Risk acceptance limit (average FAR): 10

The argument for deviating from the requirement relating to reserve buoyancy would be that there is a slight increase in average risk to personnel (according to the assumed values above), and the average risk level is in any case significantly below the acceptance limit.

In a restricted decision-making context, these two arguments are usually considered to be sufficient to conclude that the deviation may be accepted.

What can the authorities do in such circumstances? Experience has shown us that this is actually not much. They can challenge the performance of the QRA, its data, assumptions and simplifications. But if these challenges do not change the results dramatically, there is not much the authorities can do under a functional, risk based legislation, as is the case in UK and Norway.

In Norway, the authorities have for some time warned against the use of risk analysis in order to "prove" that safety systems and/or functions may be removed (referred to as "misuse of risk analysis"). On the other hand, the authorities have put considerable emphasis on the use of risk analysis and risk acceptance criteria, which tends to support the view cited above. Nevertheless, it seems unsatisfactory that the authorities apparently can do little to change the conclusion, if a company decides to use risk analysis in such a manner.

In a narrow approach to decision-making purely based on arguments relating to calculated risk and economic values, the picture is typical as far as the effect of rare, but very severe accidents is concerned:

- If reserve buoyancy is installed, this represents a significant investment, which has a high probability (95–99%) over the lifetime of the installation never to be required and as such may be considered as a "wasted" investment.
- There is a small probability (< 5%) that the investment will be extremely profitable for the company as well as for the personnel on board, if an accident occurs where this function prevents capsizing.

It is often seen that the management of a company is willing to take the resulting risk of a few per cent, even if the economic penalty may be high, if the accident should occur. Some companies were unwilling to take such risks in the 1980s and 1990s, but our impression is that companies are more willing to accept such risks in today's climate. One exception is the Thunder Horse project.

We believe BP is extremely appreciative of the fact that reserve buoyancy had been provided for their Thunder Horse production installation installed in 2005 in the US Gulf of Mexico. The first media announcements by BP representatives praised the Swedish design company for providing this function in a legislative environment where this is not required.

4.3 Application of the Decision-making Framework

We believe that a broad decision-making process is essential in order to make decisions about such crucial aspects of major hazard prevention for an installation. The main elements should be as outlined in Chapter 3:

- framing of the problem
- generation and evaluation of alternatives
- managerial review and decision.

4.3.1 Framing of the Problem and Alternatives

The framework for the decision-making should be a broad process involving all relevant stakeholders, the cautionary principle as well as use of the ALARP principle.

The generation of alternatives is in the present case somewhat trivial, the options being to install the reserve buoyancy in the deck or not to install it, and there are no compromises or alternative solutions.

The evaluation of alternatives should consist of quantitative and qualitative considerations and arguments.

4.3.2 Quantitative Results

We have already presented the effects in terms of average FAR values for personnel on board; see page 95. We would also expect cost-benefit and cost-effectiveness calculations to have been made. Typically, we may have the following result:

- Expected cost (NPV) of averting a statistical life lost through installation of reserve buoyancy: 380 million NOK

We refer to the expected cost of averting loss of a statistical life as the ICAF (Implied Cost of Averting a Fatality) value. We would expect sensitivity analyses to be performed, in order to illustrate effects on quantitative results. Table 4.1 presents some illustrative results of what could come out of a sensitivity study, using ICAF to measure performance.

The sensitivity study results show considerable variation, including cases where costs of averting a statistical fatality are certainly not grossly disproportionate in relation to benefits. This applies to the two last rows of Table 4.1.

Table 4.1. Hypothetical results of sensitivity study

Variation	**Resulting ICAF value** (million NOK)
10 times higher failure frequency for ballast tank valves	85
10 times lower failure frequency for ballast tank valves	595
Increased influence of common mode failure for valves	117
10 times higher failure frequency for operator response in case of emergency involving ballast tank flooding	43
10 times higher collision frequency by passing merchant vessels	19

4.3.3 Qualitative Evaluations

The qualitative evaluations should consider aspects that are not easy to quantify, but which may be just as important as the quantitative results. Factors that should be considered include the following:

- use of good practice
- use of codes and standards
- engineering judgement
- evaluation of robustness
- stakeholder consultation
- tiered challenge.

We have already commented on some of these aspects. The provision of reserve buoyancy in the deck would be considered good practice, and is in accordance with applicable regulations from NMD.

The evaluation of robustness should be an extensive discussion. The following may serve as a brief summary of such a discussion. Reserve buoyancy in the deck is an aspect of robustness: it provides the installation with a survival capability in the case of very severe structural damage, severe damage to the ballast system or severe flooding of watertight compartments.

Experience has shown that robustness is an important aspect. The capsizing and sinking of Petrobras' P-36 floating production installation in March 2001 showed that what were supposed to be watertight doors to compartments were open and thus susceptible to flooding. In March 2004, a supply vessel rammed at almost full speed a Norwegian semi-submersible mobile drilling unit on the Norwegian Continental Shelf. The force of the collision was quite high, due to the vessels' speed. If the installation had been 20 years old, the column in question might have been lost and the unit dependent on the reserve buoyancy. In the actual case the unit was quite new with a high structural capacity, and only minor damage resulted. Common to both these two incidents is the fact that they occurred because of mechanisms that usually show very low contributions to personnel risk in QRA studies. This may also be stated for the mechanism of the loss of "Ocean Ranger", as described in Section 4.1.1.

Stakeholders relevant to the decision on reserve buoyancy in the deck are the following:

- the operator (oil company)
- partners in the field
- contractors and suppliers
- employees of oil company and contractors
- authorities.

Consultation with some stakeholders is a challenge at an early stage of the concept selection stage. There are usually no employees available, few or no contractors are selected, *etc.* However, union representatives may act on behalf of employees.

It should be noted that the decision about reserve buoyancy in the deck must be taken at a very early stage of concept development. The possibility for the authorities to influence decision-making in the very early stage is often quite limited. The authorities normally require a recommended solution to be presented for their approval or acceptance. The role of the authorities in the phase where important concept selections are made is often that of informal influence.

4.3.4 Managerial Review and Decision

A summary of the review and decision by management as a result of a broad decision-making process could be formulated as follows:

It has been noted that the assigned fatality risk on the installation will not be excessively high even if reserve buoyancy is not provided. Whether costs of providing the extra barrier are in gross disproportion or not is somewhat of a borderline issue. They may be considered to be in gross disproportion if only expected values are considered. But this conclusion is quite sensitive to assumptions and premises in the cost-benefit analysis. If the collision frequency of passing merchant vessels is increased by a factor of 10, the extra cost is certainly not grossly disproportionate.

It is further noted that reserve buoyancy in the deck is a robust solution to all scenarios involving severe structural damage or severe ballast system failure, including flooding of ballast compartments. It has been demonstrated by a recent near-miss, where total loss, and multi-billion NOK asset loss, was prevented by this extra barrier. At least three total losses in the past (with well over 200 fatalities) could have been prevented if such reserve buoyancy had been available.

The extra barrier is certainly in accordance with authority regulations, and is as such the preferred solution from a regulatory point of view. It is further in accordance with good practice, and has been adopted by owners of mobile drilling units for more than 20 years.

It is the decision of management after a balanced evaluation of quantitative as well as qualitative aspects that reserve buoyancy in the deck structure should be provided for the floating production unit in question.

4.4 Observations

The Petroleum Safety Authority Norway commissioned a study in mid-2005 of how the offshore operating companies in the Norwegian sector have implemented the requirements relating to ALARP demonstration in the Norwegian offshore legislation. The report, see Vinnem *et al.* (2006c), has provided some useful insight.

It is commonly said that the potential for implementation of creative risk reduction approaches is often highest in the early phases of concept definition. This is obviously due to the fact that there are few limitations at that stage. Some companies have remarked, however, that the available options may be curtailed in so-called "fast track" projects, where the time for planning and engineering studies is compressed to a minimum. In these cases, there is little or no time to conduct studies and evaluations that may be used to define the additional risk reduction actions.

Another aspect of early planning and design phases is that there is often a lack of formal project management routines, to the extent that documentation of decisions and decision support is not always carried out very extensively. Decisions may be taken quickly and informally, if the solution is considered obviously a good one. Documentation of the decision-making may be lacking in these cases.

On of the companies referred to an ALARP-report for the project prepared after the project had completed the detailed engineering phase. This report contained a summary of all decisions taken in the previous phases but not documented previously.

Bibliographic Notes

The case in this chapter is based on Vinnem (2007). The discussion in Section 4.4 is based on Vinnem *et al*. (2006c).

5

Applications – Operations Phase

The main case study in this chapter is related to an existing platform which is part of a so-called "production complex" *i.e.*, with bridge linked installations. The platform in question is a production platform.

The scope of the case is a significant modification of the installation, entailing the addition of new production equipment, which will have an impact on the risk level. New equipment units will mean additional potential leak sources with respect to gas and/or oil leaks, which may cause fire and/or explosion if ignited.

The decision to be made is whether or not to install additional fire protection for the personnel in order to reduce expected consequences in the rare event of critical fires on the platform. Some of the background for the case is given in Section 1.2.6.

5.1 Decision-making Context

The operator has formulated risk acceptance criteria for the personnel on the installations, expressed as FAR values – Fatal Accident Rate values. The FAR value is defined as the expected number of fatalities in 100 million exposed hours. The relevant regulations have requirements for maximum annual impairment frequencies, for certain defined so-called "main safety functions". The aim of these functions is to protect personnel in the case of severe incidents. One of these functions is the need to provide safe escape ways from hazardous areas back to safe areas for a certain period after initiation of an incident or accident. This is called the escape ways main safety function. The maximum annual impairment frequency for main safety functions is $1 \cdot 10^{-4}$ per year.

The escape ways may be impaired by several mechanisms, *i.e.*, through physical obstructions (blocking) due to severe structural damage, as well as through temporary conditions whereby the escape ways are rendered impassable due to high heat loads and/or dense/poisonous smoke.

The resulting frequency of escape way impairment for the base case design is shown in Table 5.1, based on the risk analysis performed for the installation. The calculated impairment frequency for the installation is $3.8 \cdot 10^{-4}$ per year. This result was significantly higher than the acceptance limit, $1 \cdot 10^{-4}$ per year. But due to the fact that new regulations are not given retrospective applicability in Norway (except in very special cases, see Section 4.1.2), the current limit for impairment frequency cannot be made binding for the installation in question, which was designed several years before the current regulations came into force. Consent to operate the installation had to be given, in spite of the relatively high impairment frequency for escape ways, not in accordance with current regulations.

Table 5.1. Impairment frequencies for the main safety function "escape ways"

Case	Annual impairment frequency	Percentage change
Base case	$3.8 \cdot 10^{-4}$	
New equipment	$4.0 \cdot 10^{-4}$	5.2
Risk reduction implemented	$3.8 \cdot 10^{-4}$	−0.4
Acceptance limit	$1.0 \cdot 10^{-4}$	

The implication of the result for the base case design is that the escape ways on the platform are not well protected against heat and smoke loads in a number of fire scenarios. In the particular area, risk assessment results show that dense smoke from oil export pumps on a low level may create a severe visibility problem for the use of escape ways from almost all other areas. Since fires in oil export pumps are among the most frequent fire scenarios, this is a significant problem for the platform, which remains unsolved.

When the new equipment was planned and engineered, a 5.2% increase of the impairment frequency was the effect of introducing new potential leak sources. However, the company proposed some risk reduction measures that reduced the impairment frequency by a corresponding value, so that the resulting frequency after both new equipment and risk reduction actions showed a net reduction of 0.4% (see Table 5.1).

It should be noted that the FAR values both before and after the installation of new equipment were significantly below the risk acceptance limit for FAR values, so that these values had no influence on the decision-making.

The operator in question had as its sole goal in the present case to satisfy the risk acceptance limits. The FAR value limit was relatively relaxed and presented no challenge to the design. The operator interpreted the values for the impairment frequencies to mean that an acceptable solution would be that no further increase of the escape ways impairment frequencies should result from the adding of new process equipment. A minimum solution to the decision problem was to adopt risk reduction proposals implied by the reduced values in Table 5.1.

The Norwegian regulations require the companies to perform ALARP evaluations, but say little about how this should be implemented or documented. This means that many companies will claim that ALARP evaluations have been performed, but there is no documentation showing which evaluations have been carried out. In this case too, the company claimed that ALARP evaluations had been performed.

The authorities did not agree with the adopted solution, because it did not address the fundamental issue, but could do nothing, due to lack of any legal basis.

5.2 Deficiencies and the Need for an Alternative Process

The main deficiency of the process that the company had conducted is the lack of a broad process to seek for alternative solutions and of a comprehensive decision-making process where all relevant aspects are taken into consideration.

No emphasis in the decision-making was placed on the fact that the impairment frequency was relatively high in the base case, *i.e.*, before modifications were introduced. Nor was any consideration given to the fact that the installation in question was planned to have a long operational life. With such a long life, the escape ways problem is quite likely to become a reality some time during the operational period. A final factor is the expectation that the installation in question will have several other installations bridge connected to it, at some point, such that the high exposure of personnel to fire and smoke in an accident scenario may affect a higher number of people in the future.

The operator in the present case had taken a restricted "minimum scope" approach to the management of HES, consisting of "satisfy-risk-acceptance-criteria" with minimum resources and no investment beyond what was strictly necessary to comply with legal requirements. The opposite approach, which we could call a "proactive risk reduction approach" to management of HES would not only seek minimum solutions, but would look at costs and benefits in a wider context.

From the authorities' point of view, the outcome of the management process was quite unsatisfactory – the HES management approach should be improved.

5.3 Framing of Decision Problem and Decision Process

This section describes how the decision-making process could alternatively be conducted and how alternative results and decisions could be produced for the case outlined in Section 5.1 above, if a risk reduction perspective is adopted in line with the principles introduced in Chapter 3.

5.3.1 Goals and Criteria

The following goals of the company are formulated:

A guiding principle for the company's approach to risk acceptance is that the ALARP principle shall be implemented. Risk levels as low as reasonably practicable (ALARP) shall be achieved by the implementation of risk reducing measures (technical, operational, organisational) that comply with all the following criteria:

(a) are technically and operationally feasible

(b) have a significant risk reduction effect in relative terms, when compared with the initial risk levels, after due allowance for the additional risk associated with their implementation, operation and maintenance

(c) do not involve costs grossly disproportionate to the expected benefit.

The ALARP principle shall be applied for all relevant dimensions of risk, personnel, environment, and assets.

Furthermore, the company has a written instruction stating that the decision-making process and its results shall be documented. There is a procedure for conducting ALARP evaluations, which includes the following elements:

- Description of all identified risk reduction proposals for risk to personnel, environment and assets.
- Analysis of risk reduction proposals shall be qualitative as well as quantitative. The qualitative approaches should be:
  - use of good practice
  - use of codes and standards
  - engineering judgement
  - stakeholder consultation
  - tiered challenge.

  Cost-benefit analysis is the appropriate quantitative analysis approach, when relevant.
- Documentation of those proposals that are not decided for implementation
- Implementation plan for those risk reduction proposals that will be implemented.

5.3.2 Problem Definition

The decision problem may be defined as follows in the revised management context:

- A fairly new installation has been designed and installed with unsatisfactory protection of personnel during use of escape ways against fire and explosion effects.
- The installation is intended to have an important function at the field for a long period, as the only installation to process all oil and gas from the field, which is expected to continue to operate for the next 30–40 years. This crucial role of the installation will make the problems indicated above even more critical, due to
  - increased importance of escape ways through tie-in of new bridge connected installations in the future;
  - the possibility that the life of the installation will be even further extended in the future through installation of new equipment, whereby the manning level will also increase.

- Adding of some new equipment from time to time will increase the probability of a severe fire or explosion and thus intensify the need for improvement.
- The company management has therefore encouraged the organisation to attempt to find solutions to the problem outlined above with respect to protection of escape ways.

5.4 Generation and Assessment of Alternatives

5.4.1 Generation of Alternatives

A task force group has been appointed by management in order to find possible solutions to the problem associated with the protection of escape ways. The following alternatives have been proposed:

1. Minor improvement in order to compensate for increased risk due to new equipment, but no further reduction.
2. Installation of protective shielding on existing escape ways together with overpressure protection in order to avoid smoke ingress in to the enclosed escape ways.
3. Installation of additional escape routes with sufficient protection in order to provide redundant escape routes.

An additional option is obviously to do nothing at all: accept the situation as it is.

5.4.2 Assessment of Alternatives

The task force group has analysed these options, and provided the following information:

- Risk reduction in terms of ΔPLL (PLL – Potential Loss of Life), in relation to the base case, after implementation of the new equipment, see Table 5.2. The term ΔPLL gives the change in expected number of fatalities resulting from installation of risk reducing measures. PLL expresses the expected number of fatalities for the period considered.
- Risk reduction in terms of reduction in escape ways impairment frequency, in relation to the base case, after implementation, see Table 5.2.
- Risk increase during installation phase, in terms of ΔPLL, see Table 5.4. There is an increase during execution of modifications, and a reduction when the modifications have been completed. The ΔPLL in Table 5.4 is the "net" reduction in expected fatalities, *i.e.*, the reduction in PLL from operations, with a possible increase in PLL during installation subtracted.
- Expected cost of each alternative, see Table 5.3.

It is observed that only Option 3 has an annual impairment frequency for escape ways which is below the limit stipulated in the regulations, $1 \cdot 10^{-4}$ per year. Option 2 is close however, $1.25 \cdot 10^{-4}$ per year, which is in the order of $1 \cdot 10^{-4}$ per year.

Table 5.4 shows that Options 2 and 3 have considerable cost levels per averted statistical life lost (ICAF). If these values are considered in isolation in a quantitative context, they would usually be considered grossly disproportionate in relation to the benefits, the reduction of PLL over 40 years.

Table 5.2. Overview of key risk parameters for the decision alternatives

Options	Alternative	Annual impairment frequency (escape ways)	FAR	PLL (/yr)	ΔPLL (/yr)
0	Base case	$3.76 \cdot 10^{-4}$	4.2	0.0147	
1	Limited risk reduction	$3.75 \cdot 10^{-4}$	4.4	0.0154	−0.0007
2	Protective shielding	$1.25 \cdot 10^{-4}$	3.4	0.0118	0.0029
3	Additional escape way	$9.40 \cdot 10^{-5}$	2.5	0.0088	0.0059
4	Do nothing	$3.9 \cdot 10^{-4}$	4.8	0.0168	−0.0021

Table 5.3. Overview of expected cost parameters for the decision alternatives

Options		Investment cost (mill NOK)	Annual operating cost (mill NOK)
0	Base case	0	0
1	Limited risk reduction	2	0.05
2	Protective shielding	30	0.4
3	Additional escape way	110	0.1
4	Do nothing	0	0

Table 5.4. Overview of key risk and cost parameters for the decision alternatives

Options		NPV (40 yrs) (mill NOK)	ΔPLL (40 yrs)	ICAF E(Cost)/E(saved lives) (mill NOK)
0	Base case			
1	Limited risk reduction	2.7	0.0	(Extreme)
2	Protective shielding	35.7	0.1	315
3	Additional escape way	111.4	0.2	467
4	Do nothing			

With respect to decisions about protection of escape ways, the options considered for decision-making were as noted above:

- install limited scope improvements
- install extensive heat shielding
- install additional escape ways
- do nothing.

From a very narrow risk management point of view, the alternative "do nothing" may sometimes seem attractive, because there are no costs involved, and the calculations show that there is a 99% probability that no protection will be needed during 40 years. If insurance can cover the 1% case and no legal actions can be taken against the company, this may be seen as the best alternative. The second best options is the "limited scope improvements", because the cost is limited. This is more or less what the company in question decided in reality.

In this way of thinking, the option "extra escape ways" is the worst, because the cost is high, and the option "extensive heat shielding" is the second worst.

Now, what decision to make is a management task, and would depend on the priorities of the decision-maker. The above analysis is only the quantitative part, which does not provide sufficiently broad support for making the decision. Of equal importance are the qualitative considerations of the risk aspects and the risk reduction proposals. This corresponds to adding the following dimensions (see Section 3.3.3);

A. aspects related to the consequences
B. aspects related to the uncertainties
C. aspects related to manageability.

The point is that the above calculations express conditional probabilities and expected values $P(A|K)$ and $E[X|K)$, for some events A and unknown quantities X (A may express the occurrence of an accidental event and X may express the number of fatalities next year), given the background information and knowledge K. What we are concerned about are A and X, the actual observable quantities, but our analysis provides just some assignments P and E, which express the analysts' judgements based on K, and could deviate strongly from the observables. Key factors that could lead to such deviations need to be addressed and communicated to management, as part of the overall risk picture. Sensitivity and robustness analysis are tools that can be used to illustrate the dependence of these factors and the background information K. Some examples of such sensitivity and robustness analyses are presented and discussed below. The main aspects related to the categories A–C are:

- Given possible fire scenarios, what are the smoke and radiation impacts? Which barriers will reduce the possible consequences and avoid fatalities? How reliable and robust are these barriers? Vulnerabilities?

  With the oil export pumps being the main threat, the smoke production from fires will be very dense and poisonous. The heat loads may be limited due to the smoke, but still at such levels that personnel will be fatally injured after few seconds.

The existing escape ways (external vertical towers and external gangways) do not provide any protection of personnel, to the extent that if a fire occurs, there are no barriers in order to protect personnel.

- The analysis assigns a probability of a fire of 1% during a 40 year period. However, a fire may occur and the additional fire protection will have a considerable positive effect in protecting personnel.

  Even though the frequency of critical fires is as low as 1% during 40 years, the protection of escape ways will also help in less critical fires, which will be somewhat more likely to occur. In a period of 40 years, limited fires may have a probability of typically 50%.

- The company may implement uncertainty and safety management activities that contribute to avoiding the occurrence of hazardous situations and thus accident events. Although there is a risk (expressed by the P and E, diligent efforts are made to avoid events A and obtain desirable outcomes X). These activities are mainly related to human and organisational factors, as well as the prevailing HES culture.

  One could argue that most hydrocarbon leaks are due to manual intervention on process equipment. In theory, all non-essential personnel could be removed from all areas where effects could be experienced during the use of escape ways in a fire scenario. Management may consider, however, that this places too much restriction on the operation of the installations, so that this is not feasible in practice.

  On the issue of robustness, it should be noted that heat and smoke protection of escape ways is a passive way of protecting personnel, which does not require any mobilisation or action in an emergency. Therefore, it is usually considered to be a robust way of reducing risk, as opposed to decisions that rely on equipment to be started or management actions to be implemented and followed up, which will often have a much higher failure probability.

A sensitivity study would be a natural part of a broad decision-making process. Some hypothetical results of a sensitivity study are presented in Table 5.5.

Table 5.5. Hypothetical results of sensitivity study for additional escape way

Variation	**Resulting ICAF** (mill NOK)
Base case	467
10 times higher failure frequency for severe fire	47
2 times higher radiation level on escape ways	62
Increased (2 times) proportion of south-westerly wind direction	31
Reduced (50%) proportion of south-westerly wind direction	719

The illustrative sensitivity study results show considerable variations, which suggest that the analysis is quite sensitive to assumptions and simplifications made in the analysis of risk to personnel.

5.5 Managerial Review and Decision

Management will consider the broader decision support including both the quantitative and the qualitative input from the task force group. They will sum up the situation as follows:

- Based on the initial calculations, the additional safety investments are difficult to justify.
- Applying the cautionary principle, stating that the company should pay due attention to the uncertainties, adoption of Option 2 may be justified.

The decision that may be made by management on the basis of the process indicated in this section, may therefore be to install extensive heat shielding on existing escape ways in order to provide sufficient protection, thus reducing the impairment frequency substantially for escape ways and reducing the fatality risk. However, as stated above, this would depend on how management and the decision-maker weigh the different concerns.

5.6 Discussion

Many companies have formulated "zero vision" objectives for their HES management, implying that the long-term objective is to carry out all operations without loss or damage. Sometimes it may be difficult to see the connection between such objectives and the traditional approach to decision-making, involving a narrowly-based decision-making process with short-term cost minimisation as the driving force.

Decision-making based on generation of alternatives gives a more thorough insight into the decision problem, compared to a more mechanistic approach. The process should enable a broad assessment of potential consequences and uncertainties, such that all the main aspects relating to the outcomes of the decisions are available for the decision-makers. The difficult management decision to be taken may be illustrated as follows:

If the decision to install extra protection is taken (at a cost of about 36 million NOK), the outcome over the long residual production period (30–40 years) will be one of the following possibilities:

(a) No fire occurs at all (about 50% probability), and the protection is wasted in terms of pay-back.

(b) A limited fire (not critical fire) occurs (about 49% probability), and the protection has some advantage, thus avoiding any injuries to personnel due to fire loads on escape ways.
(c) A critical fire occurs (about 1% probability), and the protection is very valuable in terms of allowing all personnel to escape to a safe location.

Obviously, if no extra protection is installed, the scenario alternatives are the same, but the outcomes in terms of pay-back are opposite:

(a) No fire occurs (about 50% probability): no cost, no other effect.
(b) A limited fire (not critical fire) occurs (about 49% probability), the lack of protection means that some of the personnel will be injured during escape, but not fatally.
(c) A critical fire occurs (about 1% probability): the lack of protection means that more than 50 persons are prevented from escaping to a safe location, many of whom may perish.

If considered in standard economic terms only, the difficult management decision is to consider the 1% probability over a 40 year field lifetime that a severe fire will occur, with possibly up to 30 fatalities, and whether to invest about 36 million NOK in protective systems and actions in order to avoid these severe consequences.

The alternative approach gives management a much broader and informative decision basis, but will at the same time expose the decision-making ability by management much more than the traditional approach. The alternative approach may therefore not be the favoured approach by management, because it also challenges management's ability to make sound decisions. Because of management's wish to protect themselves from exposure to criticism for making the wrong decision, the traditional, "mechanistic" approach to decision-making may be preferred.

One could argue that the mechanistic approach leads to better predictability, as less judgement is involved in the decision-making. That may be so, but we do not consider this argument to be strong enough to outweigh the benefits of a broad decision-making process.

The alternative approach may lead to higher investments in risk reduction, at least seen in a short-term perspective. This may be considered negative. A case is discussed in Vinnem *et al.* (1996) whereby elements of our alternative approach were employed. It was shown that a broad decision-making process led to more than twice as many funds being allocated for implementation of risk reducing measures, than would have been the case if the traditional approach had been followed.

But an argument against the alternative approach based on fear of higher investments in risk reduction, is difficult to accept if companies are serious when they formulate "zero vision" objectives. If a "zero vision" objective is adhered to, it must inevitably be expected that extra costs will be incurred as a consequence. Otherwise the objectives should be reworded to read "zero vision as long as it doesn't cost us anything".

As the decision-making is more transparent it also allows more active participation of workforce representatives in the decision-making, which is one of the overall principles of HES management in Norwegian legislation.

Melchers (2001) claims that ALARP processes are not sufficiently transparent and do not stimulate public participation. Our argument is that use of risk acceptance criteria is considerably more obscure and will in practice put a "smoke-screen" around the results, which will never be available for public consideration and debate.

One of the requirements of the decision-making process should be that it is sufficiently documented. This is important in order to make the ALARP demonstration transparent, and enable review and assessment by other stakeholders.

5.7 Observations

The first case presented in Chapter 4 focuses on a typical situation for an installation in the operations phase, when a problem area is identified, and possible solutions will have to be considered, decided and implemented. It should be noted that the description of the decision-making context in Section 5.1 and the deficiencies in Section 5.2 reflect actual cases and considerations made by the industry. For the company in question, the need to improve its decision-making is in our opinion very real. The same is true of several other companies as well.

It is sometimes claimed that application of a wide scope decision-making process is not suitable in early development project phases, such as those outlined in Chapter 4. We do not agree, and we believe that the discussion of the primary case in the present chapter will serve to underline the fact that involvement in a broad and extensive decision-making process would be even more practicable in the operations phase, because then there is usually less time pressure than during a field development project.

The major hurdle in the operations phase (as in any other phase) is if management only focuses narrowly on costs and is unwilling (or unable) to see beyond the standard calculations. In our opinion, the authorities will need to use the regulations actively, in order to change any such attitudes that may exist in companies and their managements.

Bibliographic Notes

The case in Section 5.1 through 5.6 is based on Vinnem and Aven (2006).

6

Applications – Choice of Disposal Alternative

Decommissioning may in several ways be compared to an engineering and construction project, or in this case deconstruction. But there are some very important differences which will affect the management of HES. The focus in the chapter is on the choice of disposal alternative in the planning of the decommissioning work.

The purpose of the case study is to illustrate the application of the framework presented in Chapter 3 for applications relating to decommissioning, with the focus on the following issues:

- overview of stakeholders, decision-makers and decision principles
- characteristics of the decision problem
- selection of decision alternative (disposal option).

The documentation of decommissioning alternative selection is in the public domain (Total, 2003), and these documents have been used in constructing the case study.

6.1 Case Overview

Total is the operator of the Frigg facilities (see Figure 6.1), which have been a major source of gas production in Europe for about 25 years. The Frigg field straddles the boundary between the Norwegian and the UK Continental Shelves in the North Sea and the operation of Frigg has therefore been in accordance with both UK and Norwegian legislation since commencement of operations. Production started in September 1977 and stopped on 26 October 2004.

There is a general requirement for offshore structures to be removed when the production of the field reserves has been completed, according to the OSPAR convention (OSPAR, 1992). Structures exceeding 10,000 tons may on the other hand be left *in situ*, depending on the outcome of a so-called "OSPAR process", requiring acceptance of the proposed solution by all the countries that have signed the treaty. The process further requires recommendation by the relevant national authorities, based upon a comprehensive public hearing of the Environmental Impact Assessment and the Disposal options and plans.

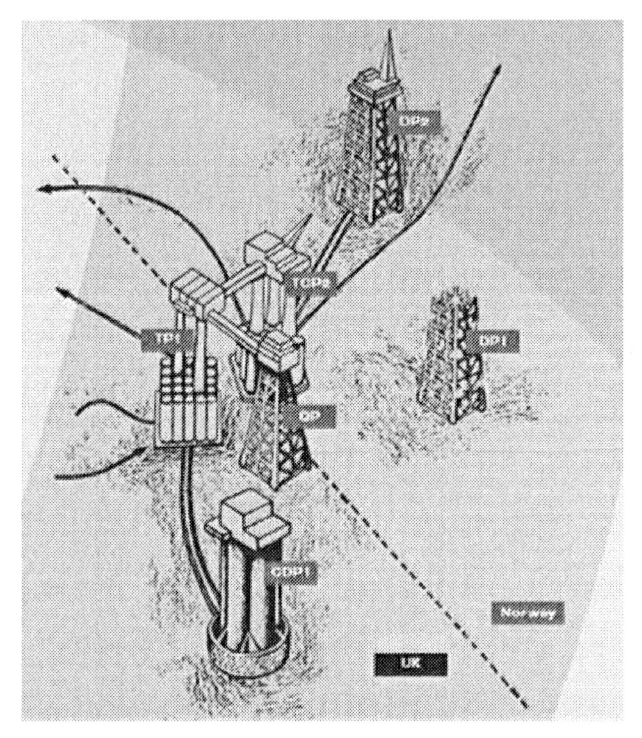

Figure 6.1. Installations on the Frigg Field (Total, 2003)

There are some minor differences between how the requirements have been implemented in UK and Norwegian legislation, and the process adopted for Frigg had to comply with both legislations. The selection of disposal options for large offshore structures is in general a complex process, and even more so in the case of the Frigg installations.

6.2 Decision-makers and Other Stakeholders

The selection of disposal options for large offshore structures (exceeding 10,000 tons) is a complex process, with many stakeholders.

On the one hand we have the operator acting on behalf of the licensees, who have been the owners of the facilities to be disposed of. The government on the other hand will usually cover 70–80% of the costs through reduced taxation. Political authorities therefore have explicitly and implicitly very strong interests in the decision-making. This is also made very clear through the OSPAR rules for decision-making, where the agreements are between states and not between private companies.

The final decision-makers in this case are the countries that have ratified the OSPAR convention, based upon recommendations from the state(s) with jurisdiction over the structures, UK and Norway in the present case.

An important group of stakeholders involves the environmentalists in various Non-governmental Organisations (NGOs). These are usually concerned about the environmental effects of the various disposal options, involving methods of removal and disposal and possible permanent effects of options where parts of the facilities are left in place.

Fishermen's organisations are usually also a strong group of stakeholders, especially when there are parts of the facilities (above sea and/or subsea) that are left in place, or dumped in the ocean.

The public at large may sometimes also become a stakeholder through the involvement of NGOs. This was clearly demonstrated in the Brent Spar case (Greenpeace, 1996), where the public was punishing the Shell company all over Europe, through different forms of actions against petrol filling stations.

6.3 Decision Principles and Strategies

The values, visions and goals, strategies and plans of the decision-maker and other stakeholders are the basis for forming the high-level decision principles and strategies, to steer the decisions in the desired direction.

The operation of Frigg has been based on the ALARP principle, in relation to acceptance and decisions about risk and risk reduction. The overall company safety objectives have been:

- to prevent accidents and thereby protect personnel, the environment and assets
- to limit the consequences of any accident
- to work for the reduction of risk over time by, *inter alia*, taking benefits from technical development and experience gained.

This has been applied to the decommissioning in the following manner. The evaluation of the various disposal alternatives has been carried out using criteria related to:

- technical feasibility
- risk to personnel
- environmental impact (including impact on society)
- cost.

The costs were expressed in year 2002 money terms and represent a 50/50 estimate, *i.e.*, an estimate with 50% probability of the cost being higher (lower) than the estimate.

Both risk for personnel and for the environment are addressed. The environmental effects may be acute in the form of impact from unsuccessful removal, or long-term impact if parts are left in place in some form.

The NGOs are sometimes not convinced by theoretical studies if they contradict the perceptions of the public or NGOs. Involvement in the thinking process and open discussion of pros and cons of different options may on the other hand be an effective approach, whereby the challenges of the decision-making process are explained thoroughly and honestly. Such a process with many NGOs in Norway and UK was conducted over a considerable period in the process leading up to the selection of disposal alternatives for the Frigg installations.

Qualitative evaluations and arguments were emphasised in the discussions with NGOs during this process, quantitative risk results were referred to where available, but the aspects given closest attention were qualitative considerations and arguments, as well as implications of the results for the different disposal alternatives.

6.4 Framing

6.4.1 Describe Goals and Objectives

Total (2003) has stated that:

"Decommissioning is an equally important part of our role as operator of the Frigg field, just as the exploration, the development and the operating phases have been. The decommissioning work will be carried out with the same high standards for health, safety and environment as during our operation of the field. The removal of the various facilities will be challenging and our main objective is to carry out the work with great attention to safety and the environment."

The overall objectives of the decommissioning of Frigg are therefore to decommission the facilities in a safe and environmentally friendly manner.

6.4.2 Problem Definition

The case considers the decision on disposal alternatives for the six offshore structures and five platform decks on the Frigg field, straddling the UK–Norwegian border in the North Sea. Three of the structures are large concrete Gravity Base Structures (GBS), all of which were installed in the second half of the 1970s, without any consideration for possible refloat and removal when decommissioned. Due to the location of Frigg, the operation as well as removal of facilities has to comply in full with Norwegian as well as UK regulations.

The need for very careful consideration and decision-making was very clearly demonstrated in the case of the disposal of the Brent Spar in the late 1990s, where the initial proposal to dump the structure was very negatively received by environmentalists and the public, much to the surprise of the Royal Dutch Shell company, who actually had the support of the UK authorities for this proposal.

Decision Criteria

The following risk-related decision criteria were established:

- Technical Risk — Based upon the risk accepted during the production phase the maximum acceptable probability of a major accident during the decommissioning operations (with the associated large financial loss) has been set as $1 \cdot 10^{-3}$ (1 in 1000). This figure is in-line with the guidelines contained in Part 1 of the "Rules for Planning and Execution of Marine Operations" published by DNV in January 1996.
- Risk to Personnel — The upper limit of tolerability for risk to personnel is $1 \cdot 10^{-3}$ per year, in an ALARP context. This criterion is in accordance with generally accepted principles applied throughout industry and supported by the UK Health and Safety Executive. For a "normal" offshore worker who spends approximately 3000 hours a year offshore, an average yearly risk of fatality of 1 in 1000 is equivalent to a Fatal Accident Rate of just above 30 (expected fatalities per 10^8 exposure hours).

6.5 Generate and Assess Alternatives

6.5.1 Generate Alternatives

The process of generating alternatives started with qualitative assessments and considerations. Different options were proposed, their feasibility was assessed and the realistic options were concluded. These were the options that were subjected to the formal assessment process in defining the disposal options to be selected. The following options were considered in the formal decision-making process:

- 5 platform topsides: Removal only option considered
- Steel jacket structures
  - 1 structure exceeding 10,000 tons: Removal only option considered
  - 1 structure not exceeding 10,000 tons: Removal only option considered
- 3 concrete gravity base structures: Options considered:
  - Leave intact *in situ*
  - Alternative use *in situ*
  - Topple over *in situ*
  - Refloat and removal
- Drill cuttings (subsea storage): Options considered:
  - Leave undisturbed *in situ*
  - Removal
- Infield pipelines: Options considered:
  - Leave water-filled *in situ*
  - Removal
- Export pipelines: Reuse only option considered

6.5.2 Selection of Method

A coarse assessment of the decision problem was performed in line with the approach in Chapter 3 in order to determine the analysis approach. Clearly in this case the expected consequences are very large (for example related to very high costs of some alternatives, the environment and safety), and the uncertainties are very large (for example due to long time horizon), and the problem needs to be classified under the "Complex decision" category.

6.5.3 Assess Alternatives

Classification of consequences, uncertainties and manageability factors are discussed in the following.

Table 6.1 and Table 6.2. at the end of the section summarise the most extreme solutions, the removal of all concrete structures and the leave *in situ* option, and the main performance measures used, as well as the main uncertainty factors.

Consequences

Below we summarise important aspects of the consequences of the decision problem, following the check list in Section 3.3.3:

(a) Potential consequences	Extensive societal costs incurred by disposal operations, possible fatalities and environmental impact if operation fails.
(b) Ubiquity	Both UK and Norway affected by costs, as well as possible fatal accidents. Environmental impact may affect parts of central North Sea.
(c) Persistency	Disposal options may have environmental effects for virtually an unforeseen period, extending a few hundred years into the future.
(d) Delay effect	Effects of some of the disposal options may not be visible until many years after the disposal project is completed.
(e) Reversibility	In the present case there will be no possibility of restoring the situation to the state before the effect of a particular disposal option.
(f) Violation of equity	There should be no or little violation in the present case: both UK and Norwegian societies and national economies have benefited from gas production on Frigg for more than 25 years, and have had substantial tax incomes. If severe environmental consequences occur, both countries may be affected on a national level. See also (g).
(g) Potential of mobilisation	It is possible to anticipate some concern in the fishing community that long-term effects may expose certain

	fishing activities severely, in disproportion to any perceived benefits from the gas production. However, experience has shown no such concerns.
(h) Performance measures	Performance measures for consequences that may develop over a few hundred years are not easy to establish.

Uncertainty Factors

The following list summarises some main aspects of the uncertainty factors associated with the decision problem:

(a) Insight into phenomena and systems:	Long-term effects of leaving *in situ* are not known at all.
(b) Complexity of technology:	The complexity of the technology being used is considered moderate.
(c) The ability to describe system performance:	System performance is adequately described for the phenomena that are well known.
(d) The level of predictability:	Moderate level of predictability applies, as the uncertainties related to the environmental impacts are relatively large.
(e) Experts' competence:	Competence of experts is good for phenomena that are well known.
(f) Experience data:	Limited experience data available for removal and leave *in situ* options for large offshore structures.
(g) Time frame:	Leave *in situ* options have several hundred years as the applicable time frame, implying substantial uncertainties.
(h) Vulnerability of system:	The system is, at least initially, not vulnerable at all to weather conditions, *etc.* However, this may change when the structure has been left for a long period.
(i) Flexibility:	The degree of flexibility is low, once the decision to leave *in situ* has been taken. All efforts will then be focused on implementing this option, and some irreversible decisions will be taken with respect to the state that the structure is left in.
(j) Level of detail:	The disposal options were analysed over a period of 4–5 years, including very detailed studies of all potential options.

Manageability Factors
If we concentrate on the disposal of the three concrete gravity base structures, the options considered for decision-making were as noted above:

- leave intact *in situ*
- alternative use *in situ*
- topple over *in situ*
- refloat and removal.

From a management of risk point of view, the alternative use *in situ* may be claimed to be the best alternative, at least in a short-term perspective. In this case, the facilities will be operated for alternative purposes, but with practices that are based on current knowledge and thus well known. The leave *in situ* alternative is the second best, except when considered in a very long perspective (several hundred years), because the structure will eventually disintegrate at some point in time.

The last two alternatives, topple over and refloat/remove are the least favourable from a management of risk point of view, the former due to unknown hazards and mechanisms, the latter due to critical systems and capabilities not being available for inspection and verification prior to commencement of hazardous operations.

Toppling over such a large structure has never been done, and will therefore be subject to large uncertainties. For the refloat alternative, vital systems (such as ballast system piping) could not be inspected, due to being installed within the water filled concrete structure at seabed level and thus not accessible after installation of the structure.

Especially for the refloat and removal option, the uncertainty relating to control of the outcome is virtually unmanageable. As indicated above, the systems required during refloat cannot be inspected and the capabilities needed during refloat and removal cannot be verified before the operations are initiated.

The decision support provided in the case with choice of disposal options may be characterised as follows:

- The decision-making process was managed and documented, including the involvement of stakeholders, rigorously and traceable.
- Ranking of alternatives was based on analysis results, and sensitivity studies were performed in order to rule out the possibility that minor changes to assumptions and/or data would alter conclusions.
- Stakeholder consultations were conducted in both countries in order to identify possible factors that had been overlooked.
- The conclusions could not be altered through minor change of weights or preferences.

Table 6.1 and Table 6.2. below summarise some of the key performance measures used and uncertainty factors for two decommissioning options, with reference to Table 3.1 in Section 3.3.3.

For the leave *in situ* option, there is one remaining hazard for the marine traffic in the vicinity, namely the chance of hitting one of the installations due to navigational failure. Table 6.3 presents the calculated probability of collision with concrete structures, if they are left in place indefinitely.

Table 6.1. Overview of main performance measures used for two decommissioning options for three concrete structures

Alternative	Risk to personnel		Environmental effects		Mission failure probability	E(cost) (mill GBP)
	E(fatalities)	P(fatal accident)	E(energy) (mill GJ)	Impact on fisheries		
Remove structures	1.1	67%	4	Moderate negative	40%	862
Leave *in situ*	0.30	26%	1	Moderate positive	0%	266

Table 6.2. Overview of classification of uncertainty factors for two decommissioning options for three concrete structures

Alternative	Complexity of technology	Complexity of organisation	Availability of information	Time frame	Reversible
Remove structures	Medium (Cat 2)	Low (Cat 1)	Low (Cat 3)	Short (Cat 1)	No (Cat 3)
Leave *in situ*	Low (Cat 1)	Low (Cat 1)	Medium (Cat 2)	Long (Cat 3)	Some (Cat 2)

Table 6.3. Probability of collision by fishing vessels and passing vessels with concrete structures left in place

Parameter	Collision by passing vessel	Collision by fishing vessel
Annual frequency of collision with one of three Frigg installations left in place	$1.8 \cdot 10^{-4}$ per year	$4.2 \cdot 10^{-5}$ per year

It should be noted that the effects on the environment are characterised by several parameters, some of which are quantitative, and some are qualitative. Only two of these are shown in Table 6.1 above. It could be added that virtually all the qualitative parameters for the environment favour the removal of the structures, whereas the quantitative environment parameters are limited to energy consumption and emissions, and obviously favour the leave in place option.

It could further be noted that one of the three structures accounted for about 85% of the total mission failure probability, due to its uncertainties with respect to the technical state of systems and aspects that would be essential during refloat and transportation to shore.

According to the rules of Section 3.3.3, the total uncertainty ranking of the removal option would be Category 3, which will also apply to the leave *in situ* option. It could be argued though that the leave *in situ* option has a marginally lower uncertainty than the removal option, based on comparison for each uncertainty factor in Table 6.2. This indicates that the manageability of uncertainty would be slightly easier for the leave *in situ* option, compared to the refloat and remove option.

In regard to the probability of mission failure (unsuccessful removal of at least one of the concrete structures), it is noted that there is a substantial probability of such failure – the assignment is 40%. There was actually an acceptance limit stated equal to 0.1% failure probability, implying that the calculated value was more than two orders of magnitude higher. The acceptance limit itself was therefore not relevant for the decision-making.

It should further be observed from Table 6.3 that the collision frequencies are quite low. Even if we consider a period of 100 years, there is only a 2% chance of a collision occurring.

6.6 Managerial Review and Decision

With respect to selection of disposal alternatives, it was decided that the concrete structures and the drill cuttings should be left in place, the export pipelines reused, and the remainder of the facilities completely removed.

It is believed that this was the first time such substantial facilities were deliberately left in place. These decisions were accepted by all stakeholders; NGOs, national authorities and the government, as well as all European OSPAR treaty countries. No critical comments or protests were heard, and the leave in place option was accepted unanimously. The decision-making process and the involvement of stakeholders have thereby proved effective.

With respect to Figure 3.3 in Section 3.3.3, all the main items were given considerable attention:

- Decision-maker: Process over several years involving several stages of studies, technical feasibility as well as risk to personnel, environment and mission failure.
- Other stakeholders: Wide variety of stakeholders involved; companies, fishermen, authorities, governments, NGOs.
- Decision principles/strategies: Consensus seeking through involvement of stakeholders.
  Several stages of recommendations to authorities and governments with stakeholder consultation.
- Decision process: Extensive studies of alternatives conducted.
- Assess alternatives: Stakeholders consulted on plans for studies as well as study results.

With respect to decision-making, it is worth noting that the very high probability of losing one structure during decommissioning turned out to be the main decisive factor.

With such a high probability of mission failure, the decision-making was easy in practice. Reaching consensus with all stakeholders also presented no problems, owing to the extensive efforts made to integrate all stakeholders in the process of determining the recommended disposal option.

It should also be noted that no trade-off decisions were required in the present case, because the most expensive removal option also had the highest risk for personnel and the highest probability of mission failure.

6.7 Observations – Decommissioning Phase

The choice of disposal alternatives is very suitable for a broad decision-making process. The regulations actually call for such a process, with involvement of a wide range of stakeholders.

The success of the process is easily demonstrated for the selection of decommissioning alternatives for Frigg. The chosen solution entails the leaving in place of three large concrete structures, after removal of external steel fittings and installation of navigation lights. This alternative was accepted by all stakeholders without strong protests, based upon the decision-making process they had been participating in, and based on the information submitted to the stakeholders.

It may be argued that the decision on selection of decommissioning alternative is a "big" decision, wide ranging and with a high number of potential stakeholders. The present case study shows that the decision-making framework in Section 3 works well for such a process. We have not shown that the process is applicable to all decision-making situations in the decommissioning phase, but our conclusion is that it does work in all major decision situations, regardless of the project phase.

Bibliographic Notes

The case in this section is based on the Frigg Cessation Plan (Total, 2003) and Aven *et al.* (2006d).

7

Applications – Risk Indicators, National Level

7.1 Background and Introduction

Health, Environment and Safety (HES) indicators for occupational injuries have been used for many years, and these were the only type of indicators used in the offshore petroleum industry for a long time. Some HES professionals claimed that no other indicators were required; the injury based indicators could perhaps be supplemented by hydrocarbon release statistics. Typical indicators that have been previously used are:

- H-value Number of Lost Time Injuries (LTI) per million manhours
- H2-value Number of personnel injuries per million manhours.

In the late 1990s many experts acknowledged that additional indicators were needed, especially for aspects relating to major hazards, and prevention of such hazards. There were several initiatives taken towards this goal, in parallel.

At the same time, there was a dispute between the parties in the Norwegian petroleum sector in the latter part of the 1990s, as briefly outlined in Section 1.2.5. There was a need to have unbiased and as far as possible, objective information about the actual conditions and developments. The authorities, the Norwegian Petroleum Directorate (NPD) at the time, now the Petroleum Safety Authority (PSA), Norway, defined a project ("the Risk Level Project"), in order to fulfil these needs.

It might be supposed that the most reliable source of information about major accidents is statistics relating to occurrence of such accidents. However, this is not a useful source in practice. Consider as an illustration the diagram in Figure 7.1, which presents all the major accidents that have occurred on production installations and mobile units in Norwegian waters since operations started in 1965. A total of 138 lives have been lost in these accidents. There were several accidents in the period 1970–1980, two accidents in the period 1981–1990, and none thereafter. Note that accidents with helicopters are not included. The accident in 1978 occurred during commissioning and start-up preparations, with about 550 persons onboard. Five persons were working in an enclosed space deep down in a concrete column, and all perished when a fire broke out in this area.

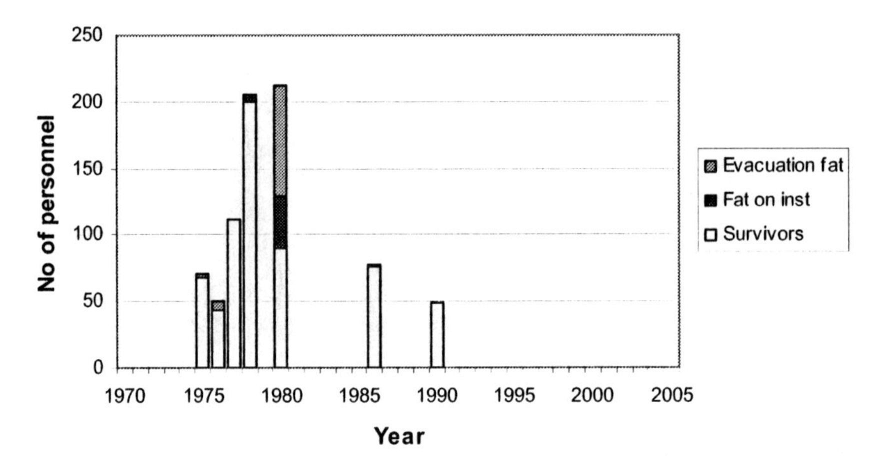

Figure 7.1. Overview of major accidents on offshore installations on the norwegian continental shelf, 1970–2005

With the low number of accidents, this source is virtually useless as input to the evaluation of major accident probability.

In fact, with regard to loss of life due to major hazards in the Norwegian offshore petroleum industry, the only events during the last 20 years are:

- Burning shallow gas blowout in 1985, with one fatality (see Figure 7.1)
- Helicopter accident in 1997, with twelve fatalities.

We therefore need an alternative approach. The Risk Level Project provides such an alternative.

7.2 Objectives of the Risk Level Project

The objectives of the Risk Level Project, usually referred to as the RNNS project, were: The PSA shall contribute to the establishment of a realistic and jointly agreed picture of trends in HES work which supports the efforts made by the PSA and the industry to improve the HES level within petroleum operations, with the following emphasis:

- Measure the impact of safety-related work in the petroleum industry, in terms of status and trends of these actions.
- Help to identify areas which are critical for safety and for which priority must be given to identifying causes in order to prevent unplanned events and situations.

- Improve the understanding of the possible causes of accidents and unplanned situations together with their relative significance in the context of risk, in order to create a reliable decision-making platform for the industry and authorities, which would enable them to direct their efforts towards preventive safety measures and emergency preparedness planning.

7.3 Overall Approach

Sometimes there is thought to be a fully objective way of expressing the risk levels through a set of indicators. This implies that expressing the "true" risk level is just a matter of finding the right indicators. However, this is a misconception. There are no single indicators capable of expressing all the relevant aspects of health, environment and safety. There will always be a need for parallel illustrations by invoking several approaches.

The basic approach adopted in the Risk Level Project from an early stage was that of triangulation, *i.e.*, to utilise several parallel paths to express status and trends of HES levels. A decision was also made to use various statistical, engineering and social science methods in order to provide a broad risk picture, covering:

- risk due to major hazards
- risk due to incidents that may represent challenges for emergency preparedness
- occupational injury risk
- occupational illness risk
- risk perception and cultural factors.

The main focus in this case presentation is on statistical indicators, perhaps so much so that our basic principles as stated above may be misinterpreted. Nevertheless, it should be stressed that triangulation and a broad basis form the fundamental approach in the project. A brief overview of the different types of indicators is given below.

7.3.1 Major Hazard Risk

The major hazard risk components for employees on offshore installations are the following:

- major hazards during stay on the installations
- major hazards associated with helicopter transportation of personnel.

A basic requirement for a risk indicator is that it is based on observations *i.e.*, we can calculate its value by a prescribed procedure using data on the performance of the system being studied. For risk associated with major hazards we have few data on losses, for example on loss of lives, and risk indicators based on other types of observations are therefore required as explained in the following. For the risk associated with major hazards on the installations, the following types of indicators have been developed:

- indicators based on occurrence of incidents and near-misses
- indicators based on performance of barriers installed to give protection against these hazards.

The incident indicators are discussed in Section 7.4, and the barrier indicators are discussed in Section 7.5. None of these indicators were readily available in the form of data already collected from operations.

The same applied to indicators for risk associated with helicopter transportation. After a search for a reliable basis, it was found that helicopter operators had all the required data registered, from which reliable indicators could be established. Separate indicators for personnel transportation by helicopter were developed in co-operation with the relevant national authority and the helicopter operators, relating to all phases of the transportation service, including taxiing, take-off, transit, approach, landing and stay on the helideck. Both transport of personnel between shore and installations as well as shuttling between installations are covered. Two categories of indicators are used:

- incident indicators
- exposure (traffic volume) indicators.

Incident indicators for helicopter transport are parallel with incident indicators for major hazards applicable to personnel who stay on the installations. Exposure indicators are used for helicopter transportation because there is a goal to keep the exposure to these risks at as low a level as possible. However, the indicators for helicopter transport are not discussed any further in this case presentation.

7.3.2 Other Indicators

The emphasis in this case presentation is on indicators for major hazards. Other hazards are also covered in the Risk Level Project, including:

- indicators based on occurrence of incidents with Emergency Preparedness Challenge
- indicators based on occurrence of occupational injuries
- indicators based on exposure of employees to selected hazards with occupational illness potential
- indicators for HES culture, based on questionnaire surveys and interviews with key stakeholders representing the different parties in the industry.

7.3.3 Leading *vs.* Lagging Indicators

It is commonly accepted that "leading" indicators are to be preferred to "lagging" indicators. Hence, there is more motivation in reporting performance of preventative measures, compared to performance in the sense of occurrence of near-misses.

The Risk Level Project has therefore also included "leading" indicators, where this has been possible and informative. The following are the "leading" indicators in the project:

- indicators based on performance of barriers that are installed in order to protect against major hazards
- indicators based on assessment of management aspects of chemical work and environment exposure
- indicators reflecting quality of operational barrier elements, based on questionnaire surveys.

The first type of indicator is discussed in Section 7.5: the two other types are not considered any further in this book.

7.4 Event-based Indicators for Major Hazard Risk

This section discusses the indicators that have been developed for major hazard risk, primarily for hazards associated with stay of personnel on the installations.

7.4.1 Indicators for Individual Hazard Categories

It is possible to draw on our prior knowledge of accidents and the factors influencing their evolution. By observing and utilising the precursors of accidents, unplanned events, faults/failures, and putting these together with our knowledge of the physical phenomena that occur (*e.g.* spills/leaks, gas dispersion, ignition, fire), we have a basis for expressing risk. This is also essentially what we do when we perform a risk analysis.

An approach based on these ideas has also been employed in the project. A number of unplanned events or situations (near-misses), referred to here as DFUs, are selected. The term "DFU" is a Norwegian abbreviation, but is often used also in English, where it could be translated as DSHA – defined situations of hazard and accident. The DFUs were selected on the basis of the following criteria:

- The DFU is an unplanned event/situation which has led, or may lead, to loss (of life and other values), and hence represents a risk contribution.
- The DFU must be an observable event/situation, and one which it is feasible to record accurately.
- The DFUs must (as far as possible) cover all situations that can lead to loss of life.

- The DFUs are important for motivation and awareness, since they are utilised in the planning and dimensioning of the emergency preparedness.

The relevant major hazards for personnel on the installation are addressed in QRA studies, and these were one of the main sources when indicators were identified. Table 7.1 shows an overview of the DFUs that have been included in the major hazard category, including the sources used. The industry has used the same categories for data registration through the database Synergi.

Data acquisition for the DFUs relating to major accidents draws partly on existing databases in the Petroleum Safety Authority (CODAM, CDRS, *etc.*), but also to a considerable degree on data collected in co-operation with operator companies, for example the database HCLIP for hydrocarbon leaks.

The method for event-based indicators is outlined above, and is further defined below. More details are described in the RNNS methodology report (SPA, 2000).

Table 7.1. Overview of DFUs and data sources

DFU no.	DFU description	Data sources
1	Non-ignited hydrocarbon leaks	HCLIP via data acquisition*
2	Ignited hydrocarbon leaks	HCLIP via data acquisition*
3	Well kicks/loss of well control	CDRS (PSA)
4	Fire/explosion in other areas, flammable liquids	Data acquisition*
5	Vessel on collision course	Data acquisition*
6	Drifting object	Data acquisition*
7	Collision with field-related vessel/installation/shuttle tanker	CODAM (PSA)
8	Structural damage to platform/stability/anchoring/positioning failure	CODAM (PSA) + industry
9	Leaking from subsea production systems/pipelines/risers/flowlines/loading buoys/loading hoses	CODAM (PSA)
10	Damage to subsea production equipment/pipeline systems/diving equipment caused by fishing gear	CODAM (PSA)

* Data acquired with the co-operation of operator companies

7.4.2 Basic Risk Analysis Model

Figure 7.2 illustrates the use of incidents and near-misses (DFUs) and their relation to the total number of anticipated fatalities, illustrated by the "number-of-fatalities-row" (*i.e.* the row of boxes indicating terminal events in the event tree, where the contributions to the number of fatalities would normally appear) at the bottom of the figure.

The diagram in Figure 7.2 has close resemblance to an event tree, and this is indeed the basis for the model. Some of the DFUs may be considered as initiating events in event trees, as they are often in quantitative risk analysis of offshore installations. This applies for instance to "unignited gas leak" (DFU1) and "[passing] vessel on collision course" (DFU5). Most of the DFUs fall into this category, with the following exceptions:

- DFU2: Ignited hydrocarbon leaks
- DFU4: Fire/explosion in other areas, flammable liquids
- DFU7: Collision with field-related vessel/installation/shuttle tanker
- DFU10: Damage to subsea production equipment/pipeline systems/ diving equipment caused by fishing gear.

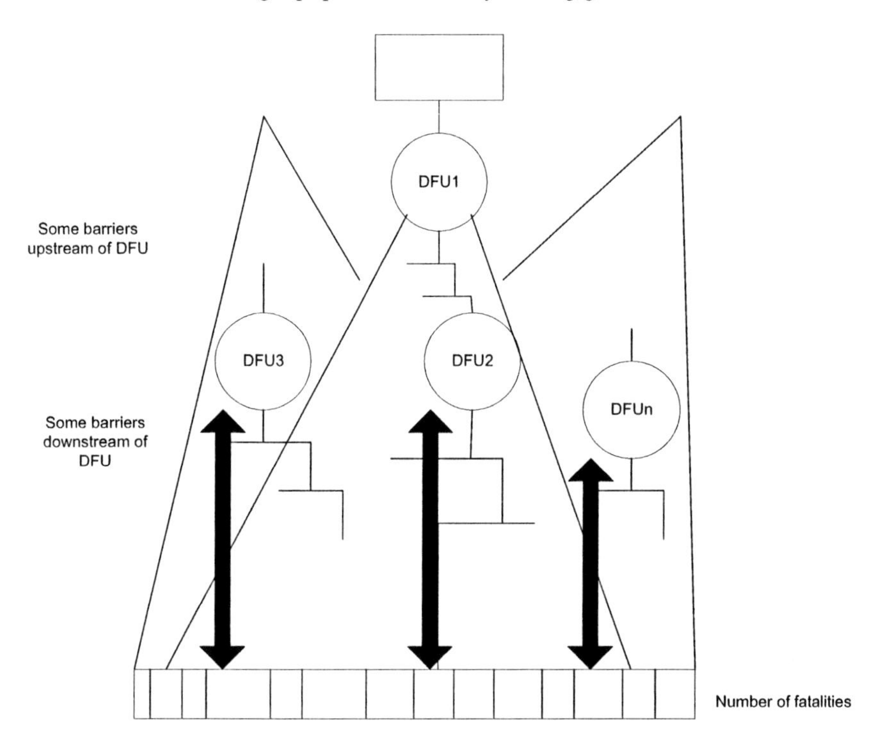

Figure 7.2. Illustration of event sequences and DFUs

These DFUs are normally not used as the initiating event in event trees, but could also be considered initiating events, if necessary.

In the diagram in Figure 7.2, the length of the arrows is intended to indicate the overall effect of the barriers, between the DFUs and the potential number of fatalities, illustrated by the event chains shown in the figure.

The DFU categories, as defined by Table 7.1, have been defined in a way that avoids overlapping, with one exception. DFU1 is unignited hydrocarbon releases (>0.1 kg/s) and DFU2 is ignited hydrocarbon releases (>0.1 kg/s). This implies that DFU2 is a subset of DFU1. So far there have been no occurrences of DFU2, so this overlap is not a troublesome one in practice. At the same time the DFU categories should cover the complete spectrum of possible outcomes from major hazards. QRA studies and extensive reviews were undertaken initially in order to ensure that no hazards were overlooked.

For the indicators that are normally considered as initiating events in an event tree, we have the following equation for the overall major hazard risk level indicator (on the installation), R:

$$R = \sum_{i}\sum_{j} NU_{ij} \cdot v_{ij} = \sum_{i}\sum_{j} NU_{ij} \cdot EX_{ij} \tag{7.1}$$

$$v_{ij} = \sum_{k}\left[\prod_{l} u_{ijkl}\right] \cdot EPLL_{ijk} \tag{7.2}$$

where the following notation is used:

NU_{ij} frequency of initiating event i, *i.e.* annual number of near-misses of category i (*i.e.* DFU type as per Table 7.1) for installation j

v_{ij} weight of category i for installation j (see list on Page 134)

EX_{ij} statistical expected number of fatalities per occurrence of an event in category i at installation $j = v_{ij}$

R indicator for annual risk level, as expected fatalities per year, given the number of near-misses, for the installation

u_{ijkl} unavailability of barrier l in accident sequence k on installation j for accident category i

$EPLL_{ijk}$ expected number of fatalities given fatal accident scenario k on installation j and initiating event gas leak (i=1).

The conditional expected number of fatalities EX_{ij}, given occurrence of the initiating event, will be a function of the performance of the consequence barriers and the potential of the accident type to cause fatalities, exactly as in normal event tree analysis.

What is special in the Risk Level Project is that EX_{ij} are not calculated based on their individual components, but overall values, referred to as "weight", calculated for typical installations, representing six broad groups, see list on page 134.

The weights v_{ij} as explained above are conditional expectations. Let us consider an unignited gas leak of a certain magnitude. This can be reformulated as follows:

$$v_{ij} = \sum_{k} P(\textit{fatality accident scenario k occurs} \mid \textit{gas leak } (i=1)) \cdot EPLL_{ijk} \tag{7.3}$$

For the DFU2, DFU4, DFU7 and DFU10, there are few or no barriers, to the extent that the weight v_{ij} will express the conditional expected number of fatalities, given occurrence of the accident, such as fire in systems not containing hydrocarbons.

The indicator for overall risk level, R, see Equation 7.1, is normalised in relation to exposure, as explained below. The final step is to transform the values into relative values, in relation to the values in year 2000. This may be expressed as follows:

$$R' = \frac{R}{V} \tag{7.4}$$

$$R'' = \frac{R'}{R'_{2000}} \tag{7.5}$$

where the following notation is used:

R' normalised value of R according to exposure, fatalities per exposure unit
R'' relative value of the normalised value R'
R'_{2000} normalised value in year 2000
V annual volume of exposure, typically manhours, number of wells drilled per year or similar

When occurrences of incidents and near-misses are considered, the extent of exposure must be observed in order to obtain a meaningful illustration of trends. If the exposure doubles, the number of incidents can also be expected to double, if the incidence rate is constant. Normalisation is the exercise of dividing the incident rate by the volume of exposure. It is therefore important how this normalisation is carried out.

There is no single measurement that is uniquely the best parameter for normalisation, and an array of parameters has to be employed. The following parameters are used in the normalisation:

- manhours (working hours)
- number of installation years (according to various types of installations, see list on page 134
- number of wells drilled (according to the type of installation where the drilling takes place).

Manhours are used as the overall normalisation parameter, not because it is suitable in all circumstances, but because a common parameter is an advantage, and normalisation against manhours has a parallel in the way risk values are often presented. Risk to personnel is often expressed as FAR values *i.e.*, the number of fatalities per 100 million manhours.

The first conditional probability of Equation 7.3 is the conditional probability of fatal outcome for accident sequence k, corresponding to terminal event k in an event tree. This means that this probability reflects all the probabilities of the functioning of protective and preventative barriers, "downstream" of the leak itself.

In principle, the majority of these probabilities could be determined from barrier data, and this is discussed further in the subsequent section. At present, these probabilities have been determined from QRA studies, where the values have been determined as average values from a number of representative QRA studies. For such purposes, the installations have been categorised into the following groups:

- fixed production installations
- floating production installations, with possible well release exposure on the installation
- floating production installations, without well exposure on the installation (wells distant)
- production complex with bridge linked installations
- normally unattended installations (NUI)
- mobile units.

The way the weights were established in the present version of the method, is as follows: for each of the six categories, some QRA studies were used as representative installations in each category. The values EX_{ij} were determined from the study results for the relevant initiating events (DFUs), and an average value of the individual values EX_{ij} was calculated. For some of the DFUs, notably structural damage (DFU8) and collision with field related traffic (DFU6), corresponding results are usually not calculated in QRA studies for offshore installations, and the weights had to be based on general accident statistics and judgements by experienced QRA personnel. Details are found in the Methodology Report (PSA, 2000).

The weights v_{ij} to a large extent implicitly reflect the performance of barrier elements, and although barrier indicators were not established from the start, data for some selected barriers were collected in the period 2002–2004. It has been considered whether the weights should be directly reflecting the barrier elements' performance data, at least partly. The weights would then in principle be calculated by Equation 7.2. The unavailability of barrier elements is multiplied following the different paths of the event tree.

This approach has not been implemented so far, because the present version of the method is not considered to be sufficiently detailed. An approach like this would require specific weights and assessments for each installation, but in the present approach, all installations are averaged into six large categories as mentioned above. In principle, it would not be difficult to expand the method to make specific assessments for each installation, but it would be quite resource demanding and thus expensive. So far, it has not been found cost-effective to take on this expansion.

Figure 7.3 shows the trend for hydrocarbon leaks over 0.1 kg/s, normalised against installation years, for all types of production installation, from the annual report for 2004 (SPA 2005). The figure illustrates the technique used throughout the Risk Level Project to evaluating the statistical significance of trends. The normalised parameter presented in Figure 7.3 is "number of leaks > 0.1 kg/s per installation years". In the present case, this has been calculated as an average for all production installations on the Norwegian Continental Shelf, whereas in other cases it has been calculated separately for the relevant groups of installations.

The last bar in this diagram is a prediction interval (90%) for 2004, based on the average level in the preceding period, 1996–2003. The intervals have been calculated from the Poisson distribution. The basis for this method, and alternatives, are discussed in Kvaløy and Aven (2005). When the value in 2004 falls in the dark grey (lower) part of the bar, it implies that there is a statistically reduced value for 2004, compared to the average for 1996–2003.

Another example is shown in Figure 7.4, where the development of the total number of major hazard near-misses and incidents is shown from 1996, normalised in relation to manhours worked in the industry per annum.

The term "relative risk indicator" is used in the diagram when the actual value is without particular significance, whereas the focus should be on the trends. In the case of Figure 7.4, the relative value is "number of incidents per 10 million man-hours".

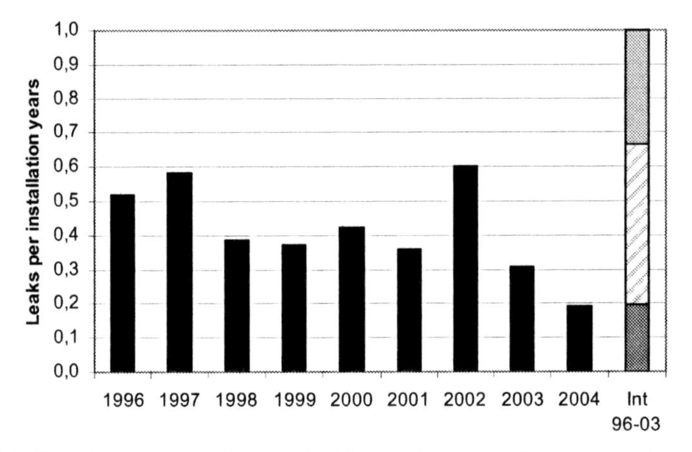

Figure 7.3. Trend, leaks, normalised against installation year, all production installations

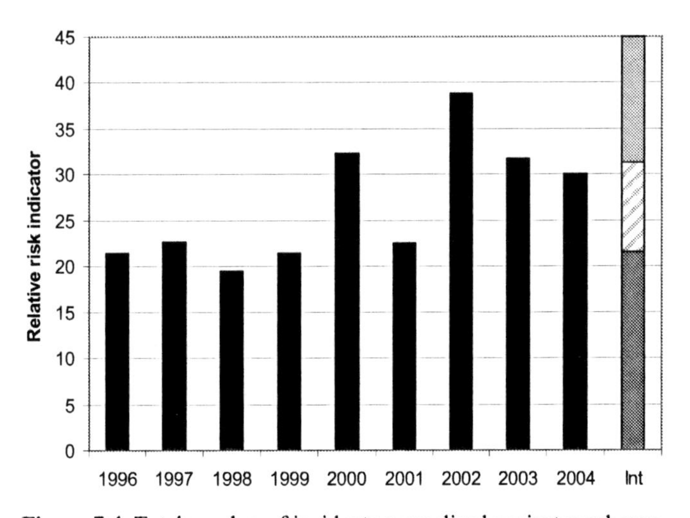

Figure 7.4. Total number of incidents normalised against manhours

One of the important observations that may be made from consideration of individual installations is that there are very distinct differences between individual installations and companies. This shows that there is clear room for improvement, a fact also underlined by Figure 7.5 which shows the average leak frequency per installation year for anonymous operator companies on the Norwegian Continental Shelf. It should be noted that the number of leaks over the nine-year period is so high that the biggest differences in Figure 7.5 are significant differences, according to the applied method. For leaks of the order 0.1–1 kg/s, there is some uncertainty related to possible, but unlikely to be significant, underreporting. However, the figure shows that those companies which generally have the highest leak frequency also have the highest frequency if we only consider leaks exceeding 1 kg/s. For these leaks, underreporting is highly unlikely.

It should be noted that Operators 8 and 9 in Figure 7.5 have very limited operations, and have not had operations for the entire period covered by the study.

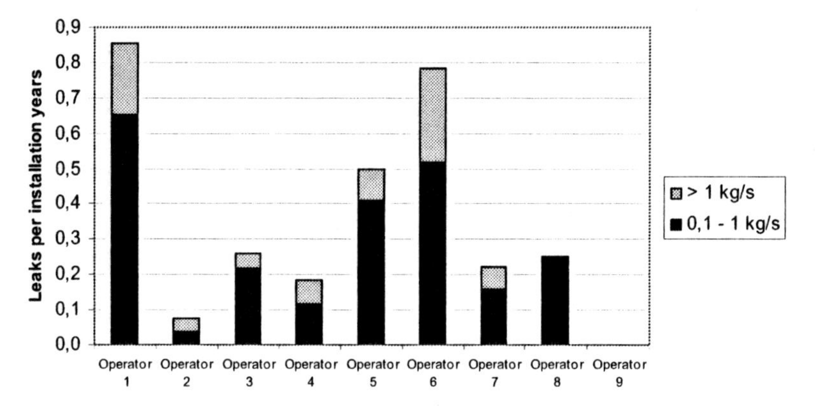

Figure 7.5. Average leak frequency per installation year, 1996–2004

The typical picture from the individual indicators is that some will show an increase, some may show a decreasing trend, and several will usually fall within the prediction interval (middle part in the diagrams), *i.e.*, so that no significant trend can be concluded. It is therefore an advantage to have an overall indicator that can balance the effects of the individual indicators, in order to identify the overall development. It might be supposed that the overall indicator would always fall within the prediction interval, but this is not the case. This is due to the large differences in weights applied to the different categories, meaning that some categories will dominate over others.

Illustrations of the overall indicator are shown in Figure 7.6 and Figure 7.7, which show the trend of the total indicator for all production installations and all mobile units, respectively. Figure 7.6 shows a significant increase in 2004, when compared to the average for the period 1996–2003. The overall impression is also an increasing trend with some variations from year to year.

Figure 7.7 shows also for mobile units a significant increase in 2004, when compared to the average for the period 1996–2003. The overall impression is also for the mobile units an increasing trend over the period. It may be observed that the variations from year to year are more extensive, compared to the production installations.

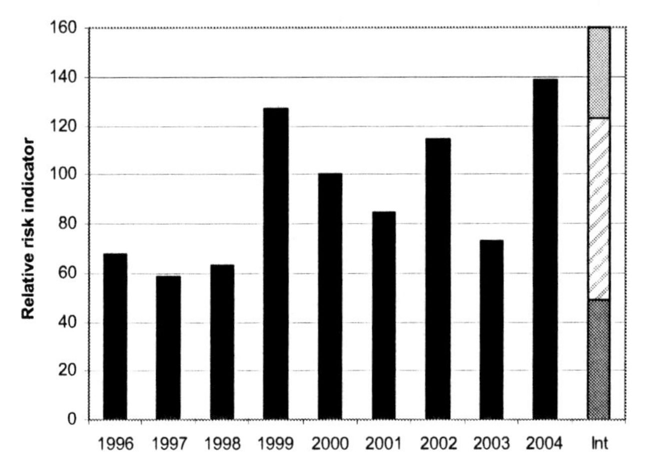

Figure 7.6. Total indicator, production installations, normalised against manhours

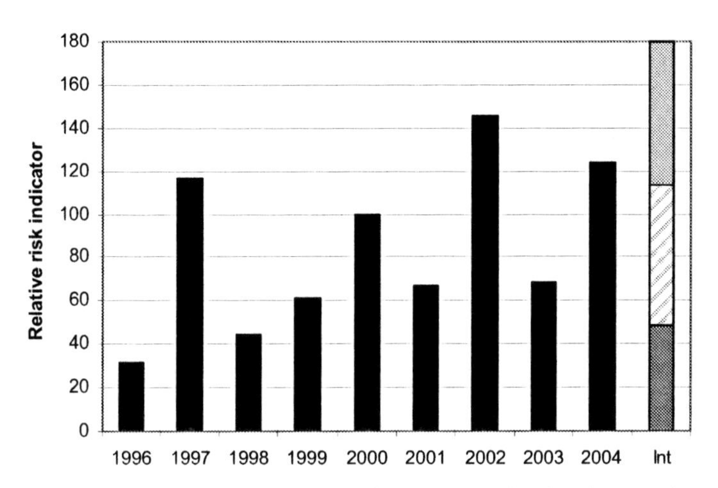

Figure 7.7. Total indicator, mobile installations, normalised against manhours

7.4.3 Challenges in the Trend Analysis

Figure 7.3 illustrated the presentation and test of trends in the Risk Level Project, by the prediction interval bar on the right-hand side in the diagram. This approach is most conveniently used to compare the last year's value with the average from previous years, but may also be used for other comparisons. There is a need to consider trends in a longer perspective. For instance, what is the long-term trend in Figure 7.3?

The long term trend in Figure 7.3 may be illustrated by comparing the average in the second half of the period 2000–2004 with the first half, 1996–1999. The average in the first half is 6.7 leaks per 10 installation years, against 5.6 for the second half. But the prediction interval increases due to the lower number of events, the lower and upper limits being respectively 4.6 and 8.7, implying that the average value for the second half is within the prediction interval based on the values for the first half of the period. The conclusion is therefore an insignificant reduction.

A more difficult question is whether the trend shown by the indicator is a representative trend or not. Consider for example Figure 7.8, which shows reported number of passing vessels (DFU5) on a potential collision course against any installation on the Norwegian Continental Shelf, the criteria for definition as potential collision course being:

- The vessel has a course which will bring it inside the safety zone (radius 500m), at a time 25 minutes prior to possible hit, and no radio contact has been established.
- The installation has mobilised the standby vessel (not including radio contact) against the incoming vessel, irrespective of distance or heading.

Fishing vessels during fishing (low speed), light crafts and pleasure boats are disregarded even though they may satisfy the criteria above, because the low energy involved in possible impacts, means they do not represent any significant hazard for the installations.

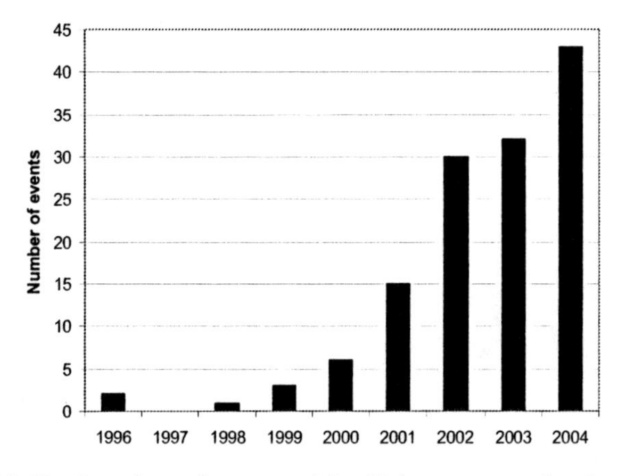

Figure 7.8. Number of vessels on potential collision course against installations

The diagram shows a sharply increasing trend after 1998. The question is whether the collision risk with passing vessel traffic has increased to the extent indicated in Figure 7.8 or not. It should be noted that two collisions with the highest energy levels occurred in 2000 and 2004, in both cases however, not due to external traffic (passing vessels) but to traffic associated directly with the offshore operations.

The important issue was that a traffic surveillance centre was started up on the Norwegian west coast in November 1998. If the trend in Figure 7.8 is plotted against the curve showing the number of installations being monitored from the centre, the two curves show quite a close fit. The traffic centre has taken over surveillance from similar operations carried out offshore, mainly on the bridge of the standby vessels. The anticipation is that the surveillance performed by the centre, which is around the clock and 365 days per year, is considerably more reliable, because the operators have surveillance as their primary task.

The interpretation of the development in Figure 7.8 is therefore that there has been, and to some extent still is, considerable under-reporting of vessels on potential collision course, because the installations have been unaware of vessels on potential collision course. The surveillance from the onshore centre is a positive step forward in order to eliminate this unawareness, and it is positive that more and more installations are being serviced by the centre, based on on-line export of radar signals directly to the centre.

There are a few other indicators that are influenced by factors that may disturb the visible trends somewhat, but none as strong as in Figure 7.8. For the majority of the indicators, this is not a problem. It is nevertheless important to be fully aware of these effects, in the discussion of interpretations and conclusions.

In the case of the passing vessels on collision course (DFU5), a new indicator was implemented in 2004, whereby the number of observed vessels on potential collision course (Figure 7.8) is normalised according to the number of installations monitored from the onshore traffic surveillance centre. The new indicator is shown in Figure 7.9.

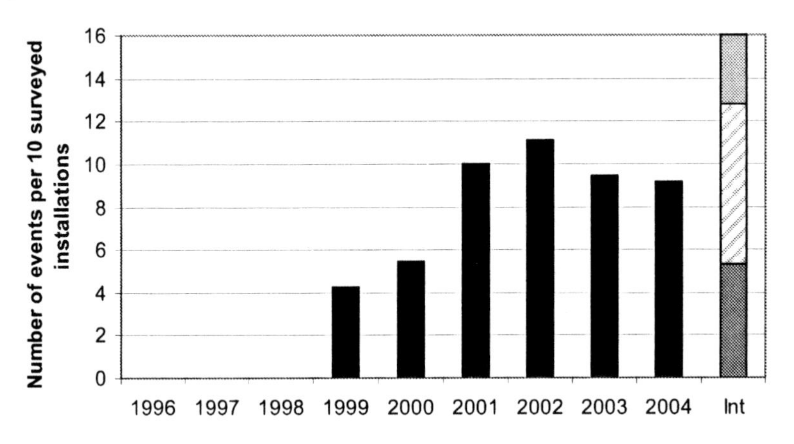

Figure 7.9. Number of vessels on collision course in relation to number of surveyed installations

It may be observed from Figure 7.9 that starting with 2001, the level of the new indicator is quite stable. Other evaluations have concluded the same result, that the level is quite stable. This may be taken as an indication that the new indicator gives a reasonable description of the risk level associated with passing vessels on potential collision course.

7.5 Barrier Indicators for Major Hazard Risk

Event-based indicators are lagging indicators which reflect experience in the past. Leading indicators are proactive indicators, and as such are often preferred. Barrier indicators may be considered leading indicators, and have as such been given high priority in the RNNS project. They were not developed from the start of the project, but were developed once the collection of event-based indicators had been firmly established. These indicators may be considered leading indicators in relation to the occurrence of future events (accidents and incidents), but will at the same time be lagging indicators in relation to the performance of barriers. However, in the context of the risk level project, we consider these as leading indicators.

The main emphasis has been placed on barrier elements that are associated with prevention of fire and explosion, but also structural barriers are addressed to some extent.

7.5.1 Barrier Elements and Performance Requirements

The terminology proposed by a working group from 'Working together for safety' has been adopted; see for example Vinnem (2006):

- barrier function
- barrier element (or system)
- barrier influencing factor.

The barrier function may for instance be "prevention of ignition", which may be divided into sub-functions: gas detection; electrical isolation as well as equipment explosion protection. One of the barrier elements in the gas detection sub-function is a gas detector, while another may be the process area operator. If we take the process operator as the barrier element, there may be several barrier influencing factors, such as working environment, competence, awareness and safety culture.

The importance of an individual barrier element is dependent on the configuration, which may be illustrated by the following examples:

- The importance of failure of an individual fire or gas detector is dependent on to what extent the remaining detectors cover the entire area.
- Downtime of one fire supply pump will affect the overall barrier function differently, depending on the extent of redundancy in the fire water supply.

The term "barrier" is not given a strict definition, but is used as a general expression. The PSA regulations require the following aspects of barrier performance to be addressed:

- reliability/availability
- effectiveness/capacity
- robustness (antonym vulnerability).

The reliability/availability is the only aspect of performance which varies significantly during operations, effectiveness/capacity and robustness being mainly influenced during engineering and design.

The following are aspects that influence reliability and availability of technical barrier systems:

- preventive and corrective maintenance
- inspection and test programme
- management and administrative aspects.

Starting with 2002, the operators of Norwegian oil and gas installations were asked to carry out two activities related to barriers:

1. Report results from testing of specified barrier elements (components), in terms of number of tests and number of faults revealed through tests.
2. Perform evaluation of overall performance of main barriers against major accidents.

Activity 1 was limited to a few barrier elements, and Activity 2 was intended to compensate for this, as it was expected to cover all aspects of main barriers. The results from 2002 were very promising with respect to Activity 1, whereas Activity 2 did not, with one or two exceptions, produce evaluations that covered overall aspects of barrier performance. The reasons for this failure are unknown. One may speculate that few companies had the required overview themselves needed to produce such an overall evaluation.

Activity 1 was continued in 2003 and 2004, with two new components included in 2004. The number of installations from which data are reported was also considerably increased, and more complete reporting from each installation was achieved. Activity 2 has not been continued, and has been replaced by other evaluations, as described in the following.

7.5.2 Follow-up of Performance by the Industry

Data collection has meant increased focus on barrier elements, requiring critical aspects to be identified at an earlier stage, components with a high failure frequency to be focused on, *etc*. Weaknesses which may affect several installations may also be identified, and trends may be spotted. These possibilities will improve further as more data are registered.

The observed failure frequency, together with a criticality evaluation, will be a basis for prioritising the maintenance work and optimisation of test intervals.

The majority of the installations are covered with respect to reporting of safety critical failures through a comprehensive data management system with several administrative functions, including reporting of data from operations and maintenance. This system also includes reporting of failures during inspection of safety critical systems, and does therefore cover a wide spectrum of barrier elements. Other companies have different solutions to maintenance and operations management. It should further be noted that this reporting does not comply with the requirements of reliability data collection and reporting, according to ISO 14224:1999 (ISO, 1999). This also means that what are considered "critical failures" in the risk level project are not defined according to that same standard but as failures likely to cause or contribute to a major accident.

Some companies have developed a system which is limited to the needs of the data collection required by the Risk Level Project. The disadvantage of such a scheme is that only a limited number of components are focused on, and the impact within the company is more limited than that which an integrated maintenance management system will have. All companies have indicated on the other hand, that they will cover all safety critical components in the system in the long run.

Successful implementation of this scheme requires that some of the challenges are overcome. The main challenges are:

- Precisely defined failure criteria and coding of failures in accordance with these. Failure to comply with definition of failures may result in too high or too low failure frequencies. One example is as follows: if the failure definition for an isolation valve is failure to close on demand, then internal leaks in closed condition shall not be recorded as failure. If internal leaks are recorded as failure, then the observed failure frequency may be too high. This example illustrates further that there may be more than one safety critical failure mode. Internal leaking in an isolation valve is indeed a critical failure mode in itself, but must not be confused with failure to close on demand.
- Different test methods may influence the occurrence of test failures. Some differences are known in relation to test gas used for testing of gas detectors. Other examples relate to wind effect during testing. Different test methods may also have differences with respect to how extensively the required function is covered.
- Recording of the number of tests is just as important as the number of failures. Some companies have simplified the issue by using the planned number of tests as an alternative to the actual number of tests.
- "Self-tests" are not included in the tests. Usually, manual tests are performed in addition, but these are not always recorded. Some companies do not carry out manual tests, as they regard the self-tests as sufficient, and have no data to report.
- The Risk Level Project does not require differentiation between components *e.g.*, fire detectors, however, most companies do differentiate between these (smoke, heat, flame).
- By and large, recording is based on planned (preventive maintenance) tests. What could be referred to as "real tests" occur in the case of actual gas leaks, false alarms, planned shutdowns and pressure relief cases. These "real tests" are in some respects more informative, because they involve parts of the systems that may be virtually impossible to test in a maintenance programme. But these "real tests" are not systematically recorded with respect to performance, as an industry standard. However, some companies have good documentation of these cases.

7.5.3 Availability Data for Individual Barrier Elements

The main emphasis is on barriers against leaks in the process area, comprising the following barrier functions:

- barrier function designed to maintain integrity of the process system (covered largely by reporting of leaks as an event-based indicator)
- barrier function designed to prevent ignition
- barrier function designed to reduce cloud and spill size
- barrier function designed to prevent escalation
- barrier function designed to prevent fatalities.

The different barriers consist of a number of coordinated barrier systems or elements. For example a leak must be detected prior to isolation of ignition sources and initiation of ESD.

Figure 7.10 shows the fraction of failures during tests. These test data are based on reports from all nine production operators on the Norwegian Continental Shelf. In 2002 great variation was observed in the number of registered failures and the number of tests. This was a “teething problem” which has not been correspondingly evident in the data for 2003 and 2004.

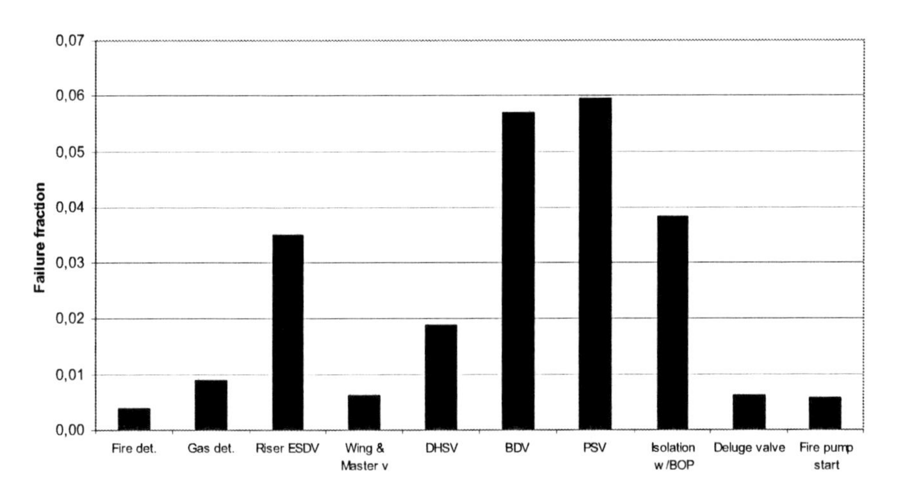

Figure 7.10. Fraction of failures for selected barrier elements, 2004

The ratio between the number of test failures and tests is an explicit expression of the on demand unavailability for the component in question, which also will reflect “environmental” aspects that influence the performance of the component, such as management and human aspects of the maintenance work. However, the calculated on demand unavailability does not take failures not detected by functional tests into account.

The relationship between the unavailability q, the observed failures X, and the completed number of tests N, is as follows:

$$q = \frac{X}{2N}$$

Trends and prediction interval are shown in Figure 7.11, calculated as explained for Figure 7.3. The number of tests is sufficiently high for most of the barrier elements, so that the interval is relatively limited, with one or two exceptions. The interval in Figure 7.11 is one of the widest, due to a relatively low number of tests for these valves, of which there are typically one to five on each installation.

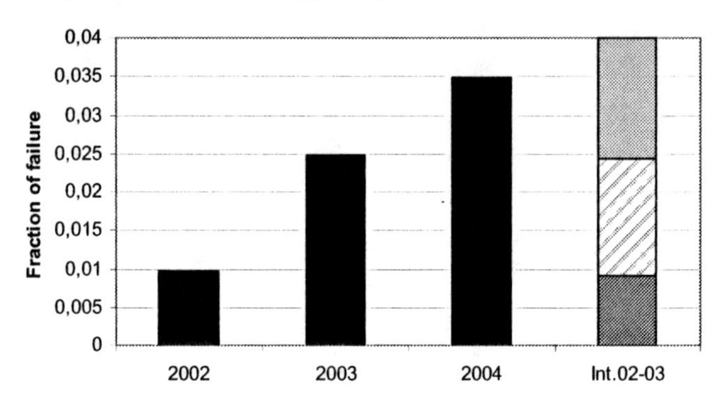

Figure 7.11. Trend for fraction of failures for riser ESD valves

In general, the relative number of failures can be claimed to be on the same level as the industry's specifications for new installations, with the exception that not all potential failure sources are tested.

Figure 7.12 shows an overview of data from muster drills, where the total number of drills and the percentage meeting efficiency requirements are given.

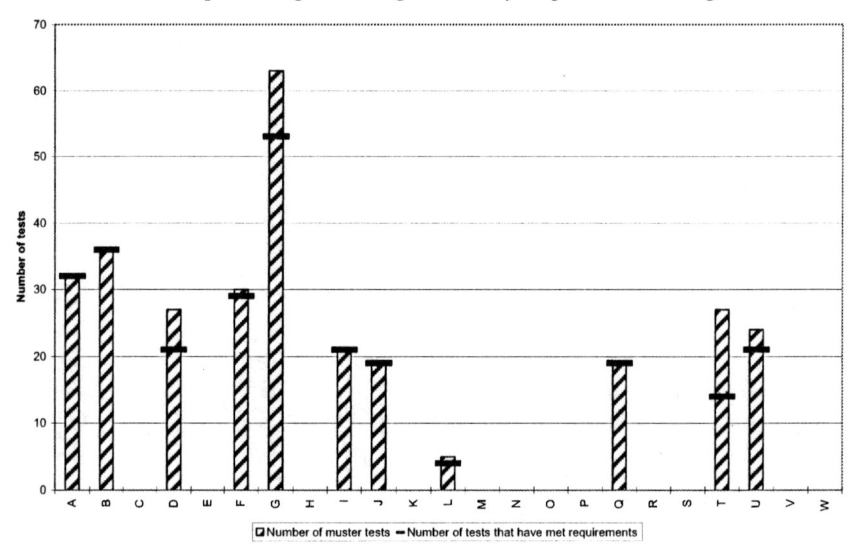

Figure 7.12. Total number of drills and number of drills meeting requirements

The efficiency requirement is usually given as the maximum time until mustering of personnel is complete and everyone is accounted for. Each installation is responsible for defining the upper limit, according to its needs and capabilities. This is then the time against which the judgement of success of drills is made. If we look at installation "B" in the diagram, 36 drills have been conducted, and all 36 have met the efficiency requirement, *i.e.*, within the maximum time allowed.

Most companies have a high percentage of drills which meet these requirements. But not all, consider for example installation "T" in the diagram, for which only half the drills have satisfied the time limit. Those with the lowest percentage are consistently those with the most stringent requirements for efficiency.

7.5.4 Overall Assessment of Barrier Performance

The initial intention was that the companies should perform an internal evaluation of overall performance of all barrier elements. When this was unsuccessful, it was replaced by an assessment performed by the project team. The general evaluation of barriers is based on reported data, meetings with some of the production operators, and experience from Petroleum Safety Authority inspections/audits. During inspections, great variation in overview of barriers and follow-up systems was noted in operating companies. This can also be seen from the general barrier indicator presented in the main report, PSA (2005):

- Certain installations on the Norwegian Continental Shelf have a consistently higher number of failures during tests of barrier elements (including the integrity barrier) than the average for the Norwegian Continental Shelf, and even more so in relation to the best installations on the Norwegian Continental Shelf.
- When the percentage number of failures during tests is compared with the average for companies, most companies fall near the average. Two companies have a markedly lower percentage of failures than the average. Both have relatively few installations.
- One company has a significantly higher percentage of failures in tests of barrier elements (including the integrity barrier) than the average for the Norwegian Continental Shelf. There is no reason to believe that this is due to lack of diligence in collecting data from barrier elements, but rather it would seem to reflect a lack of attention to following up the status and trends of barrier performance, possibly also indicating too long test intervals.

Figure 7.13 presents the relative barrier indicator developed to reflect overall barrier performance.

The bars in the diagram have their starting points at the top. The different barrier functions are shown with separate shading; containment; fire detection; gas detection; isolation; fire fighting; mustering. The depth of the contribution from a certain barrier function corresponds to how good the performance of that barrier function has been, based on the reported data. The contributions from each barrier function are then added, to give the total of all barrier elements considered.

Consider for instance installation "A", where virtually all of the barrier functions (with separate shading) are deeper than the average. This implies that the fraction of failures in testing have very low values, for virtually all of the barrier functions. This means that installation "B" has the best total performance in the diagram, and installation "K" the worst total performance.

Each installation on the Norwegian Continental Shelf for which there was barrier data available is included anonymously in the diagram presented in the RNNS project report. Only a subset of the installations is presented in Figure 7.13 in order to improve readability. The average for all installations is also shown.

When interpreting the results of this exercise in Figure 7.13 it is important to remember the limitations that apply to this indicator:

- Only a limited number of barrier elements are included.
- For those elements that are included, only the parts that can easily be tested are included.
- The actual configuration, including redundancies, capacities, *etc.* is not included.
- Only one of the performance parameters is included, the availability on demand when tested, whereas robustness and effectiveness/capacity are not included.

There is not an unambiguous relationship between the performance of the barrier function and the reliability of one or several of the barrier elements that are components of that barrier function. One should therefore be careful when ranking installations solely on the basis of the indicators shown in Figure 7.13. The other aspects of performance do not vary considerably during operation, as discussed above. The trend in availability on demand is therefore suitable for illustrating the development of performance during the operational phase.

7.6 Observations – Indicators used on National Level

The Risk Level Project has basically adopted a triangulation approach, as briefly mentioned above. A success factor for the project is to involve relevant parties participating in the industry during all phases of the project, with a view to establishing a process based on confidence in methods and quality of the results. Potential conflicts with one or several parties could distract from the focus on continuous improvement of the HES level.

The Risk Level Project is executed by PSA staff and consultants. Involvement of relevant parties is secured by involvement of the "Safety Forum" which consists of representatives from worker unions, employer representatives and the authorities. Operator companies and helicopter operators provided data and information on activities. The project has established a "HES expert group" consisting of selected specialists. The expert group plays an important part in the process of method development.

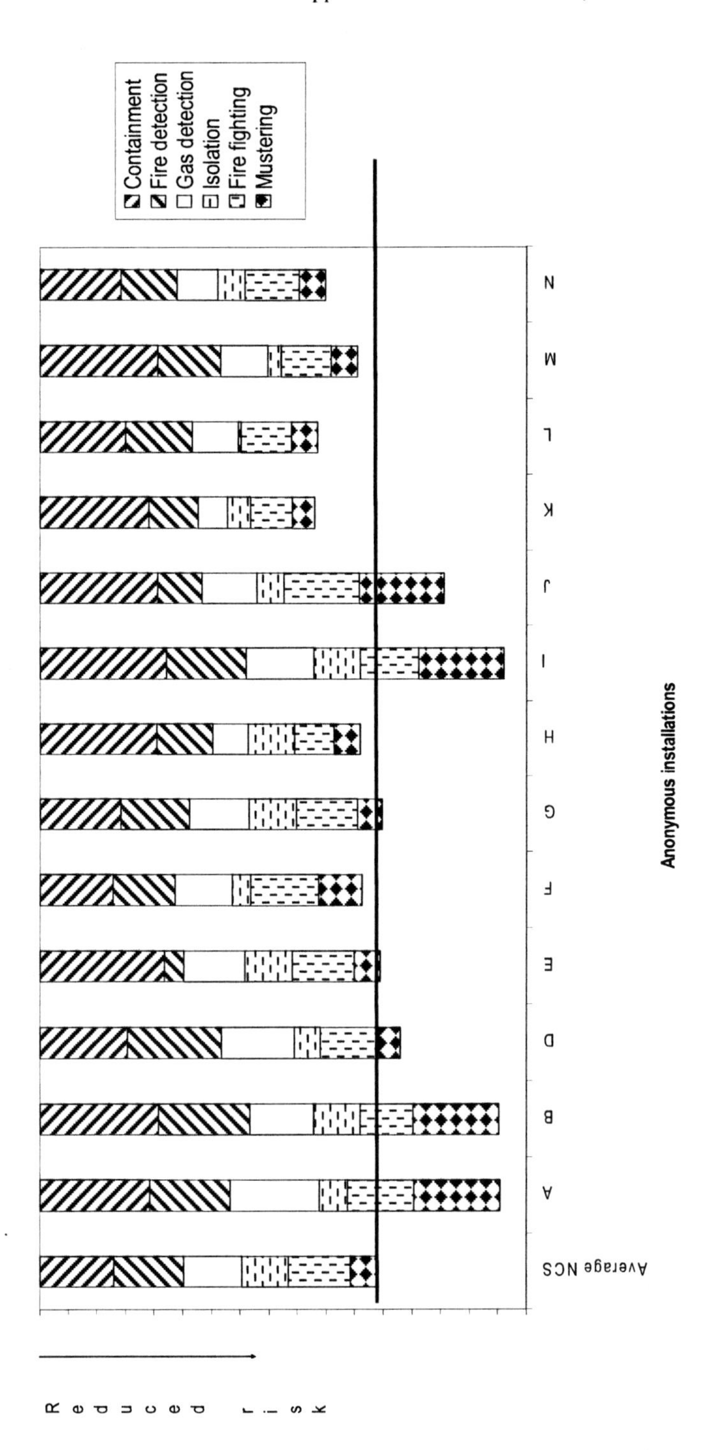

Figure 7.13. Relative barrier indicator for anonymous installations

The possibilities for integration of results from incident and barrier data analyses with the results from the analysis of the questionnaire data have so far not been exhaustively researched. A limited attempt was made after the first questionnaire survey to identify possible correlations between the statistical data on risk for major accidents and the perception of the workforce as to the relative ranking of the different major hazards. It was impossible to identify any correlations at all; they appeared to be totally uncorrelated.

It is believed that more interesting results could be obtained if the individual installations were analysed for possible correlations, for instance relating to root causes of accidents. This has so far not been done, and represents a potential future extension of the work.

Bibliographic Notes

Some of the basis for the work in this chapter may be found in Øien and Sklet (2001), Øien (2001) and Vinnem (2000). The main source is Vinnem *et al.* (2006a).

8

The Success Factors – Discussion

In this chapter we will discuss the approaches and framework introduced and used in the previous chapters. What are the main challenges? What are the key success factors?

Obviously, there are a number of factors and features of the approaches and framework that are important for risk management and the risk management process, and some of these have already been thoroughly discussed in the previous chapters. However, we see the need to summarise the main points and extend the discussion on certain topics. Two main areas are highlighted:

1. The understanding of the basic building blocks of risk analysis and risk management.
2. Implementing the framework and an ALARP regime.

8.1 Understanding the Basic Building Blocks of Risk Analysis and Risk Management

The following issues are discussed:

- Basic concepts. Uncertainty
- Assessments of alternatives
- Cost-benefit analyses and HES
- Decision principles and strategies
- Research challenges.

8.1.1 Basic Concepts and Theories – Uncertainty

Many risk analysts and risk managers do not understand the fundamental building blocks of risk analysis and risk management. This applies for example to the meaning of risk and uncertainty, and the understanding and use of models, including probability models. If we challenge a group of risk analyst professionals to explain what the risk analysis results tell us and what is meant by uncertainty in the numbers produced, we will be given a number of different answers and many of these we would characterise as quite unsatisfactory. We would draw the same conclusions if we asked the analysts to elaborate on, for instance, model uncertainty.

This is not acceptable – there is a pressing need for improvement. How can risk analysis successfully be applied in a decision-making context unless risk analysts know what they are doing and are able to communicate risk and uncertainty clearly? In our view, such improvements can only be achieved through a much more precise understanding of the fundamental principles of risk analysis and risk management.

It has been an aim of this book to contribute to such improvements, and in Chapter 2 we have addressed some key areas, including

- the use of expected values in risk management
- the understanding of some basic economic theories (for example the portfolio theory and cost-benefit analyses) and their use in HES management
- the meaning and use of the cautionary and precautionary principles
- the meaning of the concept risk aversion
- the use of risk acceptance criteria in HES management.

More fundamental issues related to the understanding and expression of risk and uncertainty are addressed in Appendix A, for example the distinction between the classical approach to risk and uncertainty and the Bayesian approach.

8.1.2 Assessments of Alternatives

In our risk management framework, alternatives are generated and the performance of these alternatives is assessed in order to support decision-making. The assessments follow a structure as discussed in the previous chapters and summarised in the following.

For a specified alternative, say A, we assess the consequences or effects of this alternative seen in relation to the defined attributes (HES, costs, reputation, *etc.*). Hence we first need to identify the relevant attributes (X_1, X_2, ...), and then assess the consequences of the alternative with respect to these attributes. These assessments could involve qualitative or quantitative analysis. Regardless of the level of quantification, the assessments need to consider both what the expected consequences are, as well as uncertainties related to the possible consequences. Often the uncertainties could be large. In line with the adopted perspective on risk, we recommend a structure for the assessment according to the following scheme:

1. Identify the relevant attributes (HES, costs, reputation, alignment with main concerns,)
2. What are the assigned expected consequences, *i.e.* $E[X_i]$ given the available knowledge and assumptions?

3. Are there special features of the possible consequences? In addition to assessing the consequences on the quantities X_i, some aspects of the possible consequences might need special attention. Examples may include:
    - o the temporal extension,
    - o aspects of the consequences that could cause social mobilisation, *i.e.*, violation of individual, social or cultural interests and values generating social conflicts and psychological reactions by individuals and groups who feel afflicted by the consequences.

   A system based on the scheme developed by Renn and Klinke (2002) is recommended, see Section 3.2.3.
4. Are the large uncertainties related to the underlying phenomena, and do experts have different views on critical aspects? The aim is to identify factors that could lead to consequences X_i far from the expected consequences $E[X_i]$. A system for describing and characterising the associated uncertainties is outlined in Section 3.2.3. This system reflects features such as current knowledge and understanding about the underlying phenomena and the systems being studied, the complexity of technology, the level of predictability, the experts' competence, and the vulnerability of the system. If a quantitative analysis is performed, the uncertainties are expressed by probability distributions.
5. The level of manageability during project execution. To what extent is it possible to control and reduce the uncertainties, and obtain desired outcomes? Some risks are more manageable than others, meaning that the potential for reducing the risk is larger for some risks compared to others. By proper uncertainty and safety management, we seek to obtain desirable consequences. The expected values and the probabilistic assessments performed in the risk analyses, provide predictions for the future, but the possible outcomes can be influenced. This leads to a consideration of, for example, how to run processes for reducing risks (uncertainties) and how to deal with human and organisational factors and obtain a good HES culture. Again we refer to Section 3.2.3.

Hence for each alternative and attribute we may have information covering the following points:

- predictions of attribute (*e.g.* zero fatalities)
- expected value (*e.g.* 0.1 fatalities)
- probability distribution (*e.g.* expressing probability of a "major accident")
- risk description on a "lower level" (*e.g.* prediction of number of leaks, expected number of leaks, *etc.*)
- aspects of the consequences
- uncertainty factors
- manageability factors.

These assessments provide a basis for comparing alternatives and making a decision.

Compared to standard ways of presenting risk results, this basis is much more comprehensive. In addition, sensitivity analyses and robustness analyses are to be performed. In Section 3.2.3 checklists are presented to identify relevant and important aspects of the consequences, and the uncertainty and manageability factors. In applications, a large proportion of the items in these lists do not apply, and it is essential for the communication of the risk results that irrelevant or non-informative items are not reported. The lists must be seen as checklists for generating important information, not for producing pages of unimportant material.

8.1.3 Cost-benefit Analyses and HES

A basic principle in safety management is the cautionary principle, stating that, in the face of uncertainty, *caution* should be a ruling principle. This principle is now being implemented in all industries through safety regulations and requirements, as discussed in Section 2.3.

On the other hand, HES management is based on the use of cost-benefit analyses to support decision-making on safety investments and implementation of risk reducing measures; see, the standard (NORSOK 2001 – Z013). Cost-benefit analyses means that we assign monetary values to all relevant attributes, including costs and safety, and summarise the performance of an alternative by the expected net present value, $E[NPV]$. The main principle in transformation of goods into monetary values is to find out the maximum amount society is willing to pay to obtain improved performance. Use of cost-benefit analysis is seen as a tool for obtaining efficient allocation of the resources, by identifying which potential actions are worth undertaking and in what fashion. By adopting the cost-benefit method the total level of welfare is optimised. This is the rationale for the approach. Although cost-benefit analysis was originally developed for the evaluation of public policy issues, the analysis is also used in other contexts, in particular for evaluating projects in companies. The same principles apply, but using values reflecting the decision-maker's benefits and costs, and the decision-maker's willingness to pay.

In this book we have drawn attention to these two perspectives and demonstrated that they are in fact inconsistent. Cost-benefit analyses do not give sufficient weight to uncertainties, being based on an attitude to risks and uncertainties which is risk neutral and in conflict with the use of the cautionary principle. A broader context for HES-related decision-making is required than that offered by cost-benefit analyses, and in the book we have characterised the features of such a context, through the proposed risk management framework.

The expected utility theory is a rational framework for decision-making under uncertainty. However, from a practical point of view, this theory is not easily implemented. The assignment of utility values determining all priorities between a number of attributes, is extremely difficult to carry out in practice, and, even if it were possible, it is not necessarily something that the decision-maker would like to do. We refer to Section 2.2.1.

We conclude that neither the cost-benefit analyses nor the expected utility theory can be used to provide clear guidance on safety investments in practice. Cost-benefit analyses have severe theoretical limitations – in their ability to reflect uncertainties and decision-makers' preferences, and the expected utility theory is impossible to carry out in practice in most cases because of a certain complexity.

A possible way of dealing with these problems is to adopt a more pragmatic view on the use of the analyses and theories, see Section 2.2.2. We acknowledge the limitations of the tools, and use them in a broader process where the results of the analyses are seen as just one part of the information supporting the decision-making, and the results are subject to an extensive degree of sensitivity analysis.

This kind of approach recognises the need for simplifications of the analyses and theories. Unfortunately, the results could be difficult to interpret, as the scope and validity of the methods and theories then become less clear. Nonetheless, we are in favour of this kind of pragmatic approach – it is the best we can do. However, we should not regard cost-benefit analyses and/or the expected utility theory as the main reference point of the decision-making process. The starting point and key reference is the risks involved, *i.e.*, the possible consequences and associated uncertainties, linked to the various alternatives considered. There are no methods that can prescribe an objective best way of handling these risks. We need to be cautious, at the same time as we may encourage risk taking activities. For example, we might wish to increase offshore petroleum activities in the Barents Sea at the same time as we would like to be cautious. Certainly a balance has to be made. We believe the following principles should be adopted for the risk management:

1. Implement some minimum safety requirements to protect human beings and the environment, see the fireproof example above.
2. Assess the risk, *i.e.*, the possible consequences, and the associated uncertainties. An important element here is to what extent we are able to manage the risks, the level of manageability. Professional risk assessments are performed, describing and evaluating the risks.
3. Balance the different concerns (safety, costs, *etc.*); implement risk reducing processes. Cost-benefit analyses may be used to support the decision-making.
4. Adopt managerial review and judgement. The decision support is evaluated in a broader context, taking into account additional concerns and information, as well as the assumptions and limitations of the tools used.

These principles have been thoroughly discussed in the foregoing chapters.

8.1.4 Decision Principles and Strategies

A number of initiatives have been taken to improve the quality of decisions on HES – there is an international trend. We refer to works by OECD and EU, and in particular contributions from UK and the UK Health and Safety Executive HSE (2001a). Many of the ideas, principles and methods presented and discussed in the UK HSE reports are obviously applicable also for the Norwegian risk management regime. In general terms, we would highlight the same aspects as UK HSE. If we take the regulator perspective; any policy intervention, and its enforcement, should meet the following principles which the Better Regulation Task Force devised in 1997:

- *Proportionality.* Regulators should only intervene when necessary. Remedies should be appropriate to the risk posed, and costs identified and minimised.
- *Accountability.* Regulators must be able to justify decisions, and be subject to public scrutiny.
- *Consistency.* Government rules and standards must be joined up and implemented fairly.
- *Transparency.* Regulators should be open, and keep regulations simple and user-friendly.
- *Targeting.* Regulation should be focused on the problem, and minimise side effects.

These principles are useful for measuring and improving the quality of regulation and its enforcement, setting the context for dialogue between stakeholders and government. Although these principles are developed primarily for regulators, they could also be seen as ruling principles for organisations and companies, following the international trend on risk management and HES decision-making.

However, some of the principles may be disputed. For example, the requirement that the decision support should be available and the decision traceable and transparent is problematic in general as it means documentation of which elements have been given weight in the decision. The authorities and the companies may not consider it desirable to trace all weights given to the various attributes. There will always be a trade-off between accountability, transparency and cost-effectiveness in the decision process.

To develop suitable decision principles and strategies, a number of concerns need to be understood and policies specified and implemented on how to deal with them. Examples of such concerns are ethical guidelines, political guidelines, application of the cautionary and precautionary principles and workforce involvement and consensus. The concerns refer to expectations from society, political signals, visions and ideal goals that should be used as guidelines for the development of the activities. These concerns will develop over time. However, at a certain point in time these concerns will be fixed, and would influence the decision process. It is therefore important that there should be a common understanding of what these high level concerns mean and how they should be incorporated into the decision process. Even though the main concerns are to be considered as constraints, they might be challenged as a part of the decision process. The need for continuous improvement and development and the fact that expectations from society, political signals, visions and ideal goals may to some degree be conflicting, call for careful reflection of these concerns in the decision process.

The variety of concerns may be grouped according to how they relate to the stakeholders, such as the political guiding rules and application of the precautionary principle. Other concerns are general and may be related to all kind of stakeholders as ethical issues. In the previous chapters we have discussed some of these concerns. Here we would just like to address one example, political guidance, to show how such concerns may influence the decision principles and strategies. The starting point is the regulation of petroleum activities on the Norwegian Continental Shelf.

Political guidance is how the politicians and authorities communicate their directions or provisions to the Regulator. In the Norwegian context, the Storting (the Norwegian parliament) and Cabinet give guidance to the Norwegian Petroleum Safety Authority and the industry through White Papers. Here some issues are repeated and highlighted. First, the petroleum industry is described as a leading industry which continuously invests in knowledge and improvement by learning from best practice. "A levelling or decline in HES performance is not in line with such objectives" the papers state. Second, the introduction of the "Zero-Philosophy" is seen as a milestone regarding attitudes and behaviour in the industry. Third, the obligation of implementing international rules and regulations is stated. This political guidance contributes to the determination of priorities and focal areas for the regulating authorities, both in revising and developing the regulations and in inspecting. instructing or guiding the industry.

We consider our recommended approach and framework for risk management and decision-making to be in line with these political guidelines. However, they do not prescribe how to carry out the risk management and decision-making processes, nor what decisions to make in specific situations. The difficult trade-offs that need to be done in many cases involving HES, cannot and should not be prescribed by political guidelines. We need structures for how to support decision-making, and that is what our framework does. The actual decision-making needs to reflect a number of concerns and the political aspect is just one of these.

8.1.5 Research Challenges

We see the need for research on many areas, in particular related to the development of theories and methodologies for

a. structuring risk decision problems and processes
b. analysing vulnerabilities and risks
c. managing risks and making decisions under uncertainty.

Different types of classification systems for characterising decision situations and risks are presented in the literature see Renn and Klinke (2002), Rasmussen (1997), Kristensen *et al.* (2005), Aven *et al.* (2005a) and Sandøy *et al.* (2005). Such classification systems are designed for structuring decision problems and guiding decision-makers on how to deal with the problems, reflecting different stakeholder perspectives, risk assessments results, *etc.* Classification as such is not the aim, but classification can be a point of departure for clarification of relationships and behavioural patterns, *etc.* We have applied aspects of these classification systems in our framework, but further research is needed to determine the most suitable schemes.

We believe that the Bayesian approach using subjective probabilities to express uncertainties provides a sound basis for risk analyses. With limited and partially relevant data, Bayesian inference is needed. However, Bayesian analysis is not straightforward for complex problems, see Aven (2003), and further research is required. Underlying this issue is the possible use of the "rational consensus" perspective (Cooke 1991) on uncertainties in risk management. Is it possible to obtain a "neutral" view of uncertainties that is acceptable for all stakeholders and thereby confine stakeholder discourse to questions to preference only? To what extent is it possible to balance such a "neutral" view and the purely subjectivist position – that probability is a subjective, personal construction?

An important issue is the choice of appropriate risk metrics, which is particularly important from a risk communication perspective, and for cases involving multiple stakeholders. The approach we applied in the Risk Level Project, as discussed in Section 7.1 for assessing risk on sector level, should be compared with other approaches *e.g.*, the social amplification of risk (Kasperson, 1992) and the analytical-deliberative process (Stern and Fineberg, 1996).

Another issue is the need for development of appropriate problem decomposition methods for risk and vulnerability identification and analysis (including extending the logic modelling techniques – such as Fault Tree and Event Tree to include influence diagrams), essential for capturing different dimensions of complex risk issues. The work must be seen in relation to the BORA project Aven *et al.* (2006a), which develops improved methodology for operational risk analysis including analysis of the performance of safety barriers, with respect to technical systems as well as human, operational and organisational factors.

To support the development of suitable theories and methods, there is a need for further research exploring the inter-relationships between economic theory, decision analysis and safety science:

- How and to what extent factors other than economic performance measures are and should be given weight, and how these factors can be measured and/or handled. Special focus should be on factors not directly related to the companies' core activity such as social responsibility and potential loss of goodwill.
- To what extent risk reducing measures are external effects for agents (companies and organisations), in the sense that they are beneficial for society but not necessarily for the agent in question.
- How these issues are influenced by public regulations and actions.

Some of the areas that need to be addressed relate to portfolio theory and safety management, the use of the cautionary and precautionary principles, the interactions between government and companies, and incentives in decision processes. This work will extend the research results reported in Chapter 2.

In such a context one also needs to be aware of the link between productivity growth and risk. Good risk management is generally claimed to be productivity enhancing, while the opposite is true of poorly designed schemes. If this is so, why is not risk management a higher priority both in theory and practice?

8.2 Implementation of the Framework

In practice, the framework is often implemented through an ALARP process or an "ALARP demonstration". There are legal requirements for such processes in both UK and Norwegian offshore legislation. In the UK there are also requirements for

ALARP demonstrations in other societal activities, such as rail transport of personnel.

In the UK there are requirements for ALARP demonstration in the Health and Safety at Work Act (1974). In the past, such requirements were also stated in the Safety Case Regulations (1992) (HSE, 1992), but these regulations were revised in April 2006 (HSE, 2006). This part of the revision is merely formal however; the authorities expect that Safety Cases for offshore installations will continue to include an ALARP demonstration.

Similarly, Section 9 of the Norwegian Framework Regulations (PSA, 2002) requires that risk reducing measures are implemented unless their cost is in gross disproportion to the benefits. Thus it may be argued that the legal requirements for ALARP demonstrations are quite similar in UK and Norwegian legislation.

There are however, some distinct differences between UK and Norwegian requirements. One difference is that UK regulations call explicitly for a written statement, describing the ALARP process that has been conducted. One of the main parts of the Safety Case is an ALARP demonstration.

In Norwegian regulations there is only an implicit requirement for documentation of the ALARP process. There is a general documentation requirement in Norwegian legislation, but no explicit requirement to document the ALARP process. The common practice is also that ALARP processes are not documented.

Another significant difference between UK and Norway is that UK authorities have focused considerable attention on ALARP demonstration/documentation, whereas Norwegian authorities until recently have paid little or no attention to Section 9 of the Framework Regulations. For several years, the Norwegian authorities focused mainly on the use of risk acceptance limits, sometimes in a rather mechanistic manner.

The UK authorities have also published several guidance documents, intended to provide the industry with practical assistance on how to interpret and conduct the ALARP demonstrations. The main documents issued by HSE are the following:

- Reducing Risk Protecting People – HSE's Decision-making Process (HSE, 2001a);
- "The ALARP trilogy", consisting of the following documents:
    - Principles and guidelines to assist HSE in its judgements that duty-holders have reduced risk as low as reasonably practicable (HSE, 2001b).
    - Assessing compliance with the law in individual cases and the use of good practice (HSE, 2003a).
    - Policy and Guidance on reducing risks as low as reasonably practicable in Design (HSE, 2003b).
- Guidance on ALARP for Offshore Division Inspectors making an ALARP demonstration (HSE, 2003c).
- HID's approach to "As Low As Reasonably Practicable" (ALARP) Decisions (HSE, 2002).

8.2.1 Experience with ALARP Demonstration

A recent project on behalf of PSA (Petroleum Safety Authority) has summarised some of the experience with ALARP processes (Vinnem *et al.*, 2006c), mainly in the Norwegian offshore industry, but also to some extent in the UK, the latter based on an interview with representatives from Health and Safety Executive.

One of the experiences in the UK is that it is difficult to document how well a process has been conducted. A good ALARP demonstration should be a broad decision-making process in which all relevant parties are invited to participate actively: management, employee representatives, safety representatives, discipline experts, *etc*. How this has been implemented is not always easy to document.

Another issue that has been highlighted in the UK is to what extent an ALARP demonstration can be based mainly on quantitative risk and cost-benefit analyses. Apparently, there have been some QRA specialists who wanted to "take ownership" of the ALARP demonstration, claiming that ALARP demonstration should be based mainly (if not solely) on QRA and cost-benefit studies. This is a misunderstanding.

It is documented clearly in the supporting documentation for ALARP demonstration that both qualitative and quantitative considerations should be part of an ALARP demonstration.

In Norway, our experience is largely the same, although considerably more limited. Some companies focus solely on the use of QRA as well as cost-benefit analysis as input to decision-making about risk reduction. On the other hand, some companies have presented broad decision-making processes that include a wide range of stakeholders.

Another issue that has been noted is the lack of a formal approach to the identification of potential risk reducing measures. Again, the QRA is far too narrow, and additional techniques have to be applied.

8.2.2 What are the Characteristics of a Good ALARP Demonstration?

As noted above, the ALARP demonstration should be a broad ranging decision-making process involving a wide range of stakeholders. An important implication of the way "gross disproportion" is defined is that the burden falls on the industry to demonstrate why a proposed remedial measure **should not** be implemented. This means that good reasons will have to be found why the benefit of a proposed measure is not sufficient in relation to costs and other burdens of implementation. If such reasons cannot be demonstrated, then the measure by default must be implemented.

The way the burden of proof is placed is often not appreciated fully, and companies are only concerned with showing that there are insufficient reasons why a measure should be implemented.

One company has described the following tools to be used for performing an ALARP demonstration:

- good practice
- codes and standards

- engineering judgement
- stakeholder consultation
- tiered challenge
- cost-benefit analysis.

We refer to Appendix B for an illustration of an extensive ALARP demonstration.

8.2.3 Needs in order to Improve Applications

There is no obvious need for more research in order to arrive at better risk management processes. The knowledge is available, in our view, but the weakness lies in its implementation.

Some people have argued that a standard should be developed which would provide a recommended approach to ALARP demonstration. A standard would perhaps be too rigid, but guidelines are certainly needed. With this in mind, we have provided a sample ALARP demonstration, see Appendix B.

A

Foundational Issues of Risk and Risk Analysis

There are a number of measures or indices that can be used to describe risk in practice, as discussed in Section 2.1. The many indices are a result of the different perspectives of risk to be found, but also stem from a lack of precision as to what the foundation of the risk analysis is. The point is that some measures are meaningful and consistent if the foundation of your analysis is Bayesian, in the sense that you see probability as a measure of uncertainty, seen through the eyes of the assessor, and some are meaningful and consistent if the foundation is classical or traditional, as will be explained in more detail below.

A.1 A Wide Spectrum of Risk Indices

Most people associate risk with the future, something unknown, and uncertainties. Let us look at an example. In project risk management, we are concerned about the cost of a production system, and say that we have historical cost data of 10 similar systems;

5, 6, 8, 9, 9, 12, 14, 20, 22, 25.

What is then the risk? Well, different perspectives and definitions would give different answers. And as we will see the differences are dramatic. Let C denote the future cost of the production system. Then we may define the expected value of C, $E[C]$, and the variance of C, $VAR(C)$, as well as the probability distribution of C, $P(C \leq c)$. From this starting point we may define a number of concepts that may be related to risk, for example

- $E[C]$
- $VAR(C)$
- $P(C \leq c)$
- The q-quantile of the distribution of C, *i.e.* the value a such that $P(C > a) = q$
- The combination of the possible consequences (outcomes) C and $P(C \leq c)$

- The combination of the possible consequences (outcomes) C and its related uncertainty
- Predicted value C^* of C
- Estimated value of $E[C]$, $(E[C])^*$
- Estimated value of $VAR(C)$, $(VAR(C))^*$
- Estimated value of $P(C \leq c)$, $(P(C \leq c))^*$
- Estimated value of a, a^*.

The estimation could be based on the above historical data, for example using the maximum likelihood principle, or it could be based on historical data as well as other information, using a type of Bayesian analysis.

Thus some may define risk by its mean and others by its variance. This seems peculiar. We can consider a very simple example, where there are two possible outcomes, 1 NOK and -1 million NOK, with associated probabilities of $1-p$ and p, where p equals $0.5 \cdot 10^{-6}$. Let D denote the outcome. Then the statistical expected value of D equals 0.5, *i.e.* $E[D] = 0.5$. Is this a meaningful measure of risk, or risk index? Obviously not, at least as a single measure. It does not provide as much information about the future possibilities and surprises as we would expect from a measure of risk. The expected value says that on the average in the long run the outcome would be 0.5 NOK, if we could repeat the situation analysed. Or alternatively, it just reflects the centre of gravity of the uncertainty distribution of D. But risk should be more than some average performance or this centre of gravity. It should in some way reflect the possible outcome of −1 million NOK. The natural answer would be to let risk cover the combination of the possible consequences (outcomes) and the associated probabilities, *i.e.* risk is described by the distribution of D,

D-values	Probability
1	$1-p$
− 1 million	p.

Hence the expected value, the variance and the q-quantile can be considered as supplementing risk indices, as they give information about the distribution of D. Separately, however they are not very informative as the distribution is poorly described by these indices. The variance is a measure expressing the spread of the distribution from the expected value. In the example above $VAR(D) = 0.5 \cdot 10^6$. Two possible ways of motivating the use of the variance are the following. On average, in the long run, with repeated similar situations, the outcome would be 0.5. The variance is a measure of the variation in outcomes relative to this mean. Alternatively, we may consider the expected value as the centre of gravity in the probability distribution, and the variance is a measure of spread of the mass around this centre. But why should we use the expected value as a reference? Say that the two possible outcomes in the above example are 10 000 001 NOK, and 9 million NOK, respectively, with the same probabilities as before. Then the expected value is increased by 10 million NOK, but the variance is the same. Is that reasonable for a risk index? Is it not also relevant to show where on the line the outcomes are, and not only how they relate to the expected value?

Yes, it is indeed relevant, as also shown by the following example. Should not risk be considered higher if the number of fatalities is expected to be 100 in contrast to 1, even if the variance is the same? Clearly, in a safety context that would be reasonable, and this is also reflected in the standard definition of risk used in safety, saying that risk is the combination of the possible consequences (outcomes) C and the related probability distribution $P(C \leq c)$. However, in economic contexts, the expected value is often used as a reference and then the variance is an informative measure of the probability distribution. Yet we find the practice in the economic literature of letting risk be associated with the variance as unfortunate, as this is not consistent with other definitions and it violates the intuitive idea that risk will be reduced (increased) if the outcomes are shifted to more positive (negative) values. Furthermore, there could be different values for the expectation. Either it is the centre of gravity in the uncertainty distribution assessed by someone, or it is considered an unknown parameter to be estimated through the risk analysis. To explain this in more detail, we have to clarify the perspective on risk adopted.

A.2 Classical, Relative Frequency Perspective

Adopting a classical (relative frequency) perspective, probabilities and expectations, are unknown values. It is assumed that there are true, underlying values of these quantities. A value $F(c)$ expressing the probability that C is less than or equal to c, *i.e.* $F(c)=P(C \leq c)$ is interpreted as the relative fraction of times the event '$C \leq c$' occurs if the situation analysed were hypothetically repeated an infinite number of times under similar conditions. Similarly, $E[C]$ is interpreted as the mean outcome when assuming that the situation analysed is hypothetically repeated an infinite number of times under similar conditions. In a risk analysis, the analysis group estimates this distribution and the expected value, based on analysis of "hard data", suitable models and engineering judgements. In some cases the estimation provides point estimates only, such as indicated above, for example $(E[C])^*$. However, in many cases we see that the estimation uncertainties are quantified, see discussion below.

The above perspective is based on the idea of underlying probabilistic terms $F(c)$ and $E[C]$. These terms are fictitious. They exist only as mental constructions, and do not exist in the real world. An infinite population of similar units needs to be defined to make this framework operational. As these underlying probabilities are unknown (uncertain), the perspective means that a new element of uncertainty is introduced, the true value of the probability, a value that does not exist in the real world. Returning to our production cost example, consider the probability that the cost C is greater than or equal to 1. According to this perspective, this probability is interpreted as the proportion of production systems with a cost of at least 1 when an infinite number of similar facilities is considered. This is obviously a thought experiment – in real life we have just one such system. The probability is thus not a property of the system itself, but of the population it belongs to. How should we then understand the meaning of similar systems? Does it mean the same type of buildings and equipment, the same operational procedures, the same type of personnel positions, the same type of training programmes, the same organisational philosophy, the same influence of exogenous factors, *etc*? As long as we are speaking about similarities on a "macro level" the answer is yes. But something must be different, because otherwise we would get exactly the same output result for each system, either a cost of minimum 1 or not. There must be some variation on a "micro level" to produce the variation of the output result. So we should allow for variations in the equipment quality, human behaviour, *etc*. The question is, however, to what extent we should allow for such variation. For example in human behaviour, do we specify the safety culture or the standard of the private lives of the personnel, or are these to be regarded as factors creating the variations from one system to another (often referred to as stochastic (aleatory) uncertainty)? We see that we will have a hard time specifying what the framework conditions of the "experiment" should be and what is stochastic uncertainty. In practice we seldom see such a specification carried out, as the framework conditions of the "experiment" are tacitly understood. As seen from the above example, it is not obvious how to make a proper definition of the population, and thus of the underlying probability.

Nonetheless, if we adopt the classical approach the probability has to be estimated. This could be difficult as the concept is a thought-construction. And we should address the accuracy of the estimation, relative to this underlying vaguely defined concept. This is done in some areas of applications, but in most cases without incorporating all sources of uncertainty. This is because a full uncertainty analysis would be too comprehensive, but even more important; it would reduce the message of the analysis. The estimates would be subject to such large uncertainties that the analysis would lose its power, see Aven (2003).

A.3 Alternative Bayesian Perspective

Our perspective to risk is different. We adopt a perspective based on the following simple ideas or principles:

- Focus is placed on quantities expressing states of the "world", *i.e.*, quantities of a physical reality or nature, that are unknown at the time of the analysis but will, if the system being analysed is actually implemented, acquire some value in the future, and possibly become known. We refer to these quantities as observable quantities. In the above example, the cost C is such a quantity. Other examples are number of fatalities, the occurrence of specified events, *etc*.
- The observable quantities are predicted. We predict the cost by C^*.
- Uncertainty related to the observable quantities is assessed and expressed by means of probabilities. This uncertainty is epistemic, *i.e.*, a result of lack of knowledge. In the example we assign probabilities $P(C \leq c)$, having a centre of gravity $E[C]$.

The notion observable quantity is to be interpreted as a potentially observable quantity – for example, we may not actually observe the number of injuries (suitably defined) in a process plant although it clearly expresses a state of the "world". The point is that a true number exists and if sufficient resources were made available that number could be found.

Focusing on the above principles would give a unified structure to risk analysis that is simple and in our view provides a good basis for decision-making.

To explain this perspective in more detail, we have to understand the probability concept, used in this way, to express uncertainties.

Consider the following example. If the possible outcomes are 0, 5 and 100, we may assign probability figures, say 0.89, 0.10 and 0.01, respectively, corresponding to the degree of belief or confidence we have in the different values. We may also use odds; if the probability of an event A is 0.10, the odds against A are 9:1. The reference is a certain standard such as drawing a ball from an urn. If we assign a probability of 0.10 for an event A, we compare our uncertainty of A to occur with drawing one specific ball from an urn containing 10 balls. The assignments are based on available information and knowledge; if we had sufficient information we would be able to predict with certainty the value of the quantities of interest. The quantities are unknown to us as we have lack of knowledge about how people would act, how machines would work, *etc.* System analysis and modelling would increase our knowledge and thus, we hope, reduce uncertainties. In some cases, however, the analysis and modelling could in fact increase our uncertainty about the future value of the unknown quantities. Think of a situation where the analyst is confident that a certain type of machine is to be used for future operation. However, a more detailed analysis may reveal that other types of machines are also being considered. As a consequence, the analyst's uncertainty about the future performance of the system may increase. Normally we would be far from being able to see the future with certainty, but the principle is the important issue here – uncertainties related to the future observable quantities are epistemic, that is, a result of lack of knowledge.

In the example in Section A.1 a probability p equal to $0.5 \cdot 10^{-6}$ is introduced. This probability expresses the analyst's assessment of uncertainty for D to be equal to –1 million NOK, or using other words, the confidence the analyst has in this outcome.

Historical data are also a key source of information when we adopt this perspective. To illustrate, consider the following example. An analyst group wishes to express uncertainty related to the occurrence of an event. Suppose that the observations show three "successes" out of 10. Then we obtain a probability of 0.3. This is our (*i.e.* the analyst's) assessment of uncertainty related to the occurrence of the event.

This method is appropriate when the analyst judges the observational data to be relevant for the uncertainty assessment of the event, and the number of observations is large. What is considered sufficiently large, depends on the setting. As a general guideline, we find that about 10 observations are typically enough to specify the probabilities at this level using this method, provided that not all observations are either "successes" or "failures". In this case the classical statistical procedure would give a probability equal to 1 or 0, which we would normally not find adequate for expressing our uncertainty about the event. Other procedures then have to be used, either expert judgments or a full Bayesian analysis, see Aven (2003).

In the production system example above, we may use the experience data as a starting point for assessing the uncertainties about C, the cost of the production system being analysed. The mean of the observations could be used as a prediction of C, if found relevant, and as the expected value of our uncertainty distribution of C.

Note that the probability assigned following this procedure is not an estimate of an underlying true probability as in the classical setting.

All probabilities are conditioned on the background information (and knowledge) that we have at the time we quantify our uncertainty. This information covers historical system performance data, system performance characteristics (such as policies, goals and strategies of a company, type of equipment to be used, *etc.*), knowledge about the phenomena in question (such as fire and explosions, human behaviour, *etc.*), decisions made, as well as models used to describe the world. Assumptions are an important part of this information and knowledge. We may assume, for example, in an accident risk analysis, that no major changes in the safety regulations will take place for the time period considered, the plant will be built as planned, the capacity of an emergency preparedness system will be so and so, an equipment of a certain type will be used, *etc*. These assumptions can be viewed as frame conditions of the analysis and the produced probabilities must always be seen in relation to these conditions. If one or more assumptions are dropped, this would introduce new elements of uncertainty to be reflected in the probabilities. Note, however, that this does not mean that the probabilities are uncertain. What is uncertain is the observable quantities. For example, if we have established an uncertainty distribution $p(c|d)$ over the investment cost c for a project, given a certain oil price d, it is not meaningful to talk about uncertainty of $p(c|d)$ even though d is uncertain. A specific d gives one specific probability assignment, a procedure for determining the desired probability. By opening up for uncertainty assessments in the oil price d, more uncertainty is reflected in our uncertainty distribution for c, using the law of total probability. The point is that in our framework uncertainty is related only to observable quantities, not assigned probabilities.

The basic idea that there is only one type of uncertainty is sometimes questioned. It is felt that some probabilities are easy to assign and feel sure about, others are vague and it is doubtful that the single number means anything. Should not the vagueness be specified? To provide a basis for the reply, let us remember that a probability $P(A)$ is in fact a short version of a conditional probability of A given the background information K, *i.e.* $P(A) = P(A|K)$. This means that even if we assign the same probability for two probabilities, they may be considered different as the background information is different. In some cases we may know a great deal about the process and phenomena leading to the event A, and in other cases very little, but still we assign a probability of say 0.50 in both cases. However, if we consider several similar events of the type A, *i.e.* we change the performance measure the difference in the background information will often be revealed. An illustrating example of this is given by Lindley (1985, p. 112). Thus care has to be shown when defining the performance measures and when evaluating probabilities in a decision making context. We always need to address the background information, as it provides a basis for the evaluation.

Next we offer some reflections on empirical control when probability is used as a measure of uncertainty, inspired by the discussion by Cooke (1992). A probability in our context is a measure of uncertainty related to an observable quantity C, as seen from the assessor's point of view, based on his state of knowledge. There exists no true probability. In principle an observable quantity can be measured, thus probability assignments can to some extent be compared to observations. We write "in principle" as there may be practical difficulties in performing such measurements. Of course, one observation as a basis for comparison with the assigned probability is not very informative in general, but in some cases it also possible to incorporate other relevant observations and thus give a stronger basis. "Empirical control" does not, however, apply to the probability at the time of the assignment. When conducting a risk analysis we cannot "verify" an assigned probability, as it expresses the analyst's uncertainty based on prior observations. What can be done is to review the background information used as the rationale for the assignment, but in most cases it would not be possible to explicitly document all the transformation steps from this background information to the assigned probability. We conclude that a traditional scientific methodology based on empirical control cannot and should not be applied for evaluating such probabilities.

A risk analyst is an expert on uncertainty assessments and probability assignments, so ensuring coherence in the assessments and assignments should not be a serious problem. The rules of probability calculus apply. There are a number of potential pitfalls in such assignments, but the risk analysts should be aware of them, and make use of tools to avoid them. Such tools include calibration procedures, use of references probabilities, standardisation, and more detailed modelling. We refer to Aven (2003) for further details.

For probabilities of this kind, the term "subjective probability", or related terms such as "personalistic", are often used. In our presentation we have avoided such terms as they seem to indicate that the results you present as an analyst are subjective while the results from others who adopt an alternative risk analysis approach are seen as objective. Why should we always focus on subjectivity? In fact all risk analysis approaches produce subjective risk results, the only reason for using the word "subjective" is that this is its original, historical name. We prefer to use "probability as a measure of uncertainty", and make it clear who is the assessor of the uncertainty, since this is the way we interpret a subjective probability and we avoid the word "subjective".

Our perspective means that risk comprises the two dimensions: a) *possible consequences* and b) *associated uncertainties*. As there are many facets of these dimensions, the framework offers a broad perspective on risk, reflecting for example the fact that there may be different assessments of uncertainties, as well as different views on how these uncertainties should be handled. This extends the common definition that risk is the combination of possible consequences and associated probabilities, as probability is the way we measure the uncertainties. However, we acknowledge that there are problems and limitations in using probabilities to measure uncertainties. This relates to our ability to express uncertainties in terms of probabilities, and acknowledges the fact not all relevant factors for decision-making can be revealed through reference to probabilities. Our analysis and framework for risk management and decision-making in this book are based on this perspective.

Bibliographic Notes

This appendix is based on Aven (2003), Aven and Kristensen (2005) and Aven *et al.* (2004). This first reference provides a comprehensive presentation and discussion of different perspectives on risk and risk analysis, as well as a bibliographic review of relevant works on this topic.

B

Example, ALARP Demonstration

This appendix presents an illustration of how an ALARP demonstration should be conducted and documented. The ALARP demonstration is made for the theoretical case presented in Chapter 5 of the book, see page 101. It is associated with modification of a production installation in the operations phase, and the need to determine the extent to which protection of escape ways should be provided for that installation.

Chapter 5 presents background information for this case study, which may be used as reference for the ALARP demonstration. Some of the information in Chapter 5 is repeated here, in order to increase readability.

B.1 Introduction

B.1.1 Purpose

The ALARP evaluation is carried out in accordance with the risk acceptance principles and criteria presented in Section 5.1. The ALARP evaluation is partly based on the risk analysis, whereby risk results and installation characteristics are considered explicitly, in order to decide which risk reducing proposals should be implemented.

The intention of the ALARP evaluation is that it should reflect values and priorities adopted by the company management, with respect to protection of life, environment and assets. As such, the conclusions will be reached on the basis of a broad decision-making process, involving all relevant stakeholders.

This document presents the performance of the ALARP process and the outcome, in terms of the remedial measures to be implemented as well as those that have been identified, but not to be implemented. The document also presents an evaluation of the resulting risk level and a justification why this is considered to be As Low As Reasonably Practicable.

B.1.2 Structure of Presentation

The approach adopted for the ALARP evaluation is presented in Section B.2, followed by the presentation of risk results in Section B.1.3. The identification of risk reducing proposals is given in Section B.4, whereas Section B.5 documents the evaluation of these proposals. The assessment of residual risk is discussed in Section B.8. A summary of those proposals that are recommended for implementation is presented in Section B.8.1.

B.1.3 Basic Assumptions and Limitations

This evaluation covers only one particular aspect of the installation, which has been in operation for about 10 years. All other aspects of health, safety and environment for the installation are disregarded in this ALARP evaluation. The following assumptions are used in the economic analysis:

- 7% internal interest
- 40 years field life
- 97,000 BOPD production rate
- 175 NOK/bbl oil price
- manning level–average POB: 40 persons
- no inflation nor change in oil price assumed
- social economics approach used: not considering tax, insurance or individual licensees

B.2 Approach Adopted in ALARP Process

B.2.1 Risk Acceptance Principles

The following goals of the company are formulated:

A guiding principle for the Company's approach to risk acceptance is that the ALARP principle shall be implemented. Risk levels as low as reasonably practicable (ALARP) shall be achieved by the implementation of risk reducing measures (technical, operational, organisational) that comply with all the following criteria:

(a) are technically and operationally feasible

(b) have a significant risk reduction effect in relative terms, when compared with the initial risk levels, after due allowance for the additional risk associated with their implementation, operation and maintenance

(c) do not involve costs grossly disproportionate to the expected benefit.

The ALARP principle shall be applied for all relevant dimensions of risk, personnel, environment, and assets.

Furthermore; the company has a written instruction stating that the decision-making process and its results must be documented. There is a procedure for conducting ALARP evaluations, which includes the following elements:

- Description of all identified risk reduction proposals for risk to personnel, environment and assets.
- Analysis of risk reduction proposals must be qualitative as well as quantitative. The qualitative approaches should be:
  - Use of good practice
  - Use of codes and standards
  - Engineering judgement
  - Stakeholder consultation
  - Tiered challenge.

  Cost-benefit/cost effectiveness analysis is the appropriate quantitative analysis approach, when relevant.
- Documentation of those proposals that are not decided for implementation and the associated residual risk level.
- Implementation plan for those risk reduction proposals that will be implemented.

B.2.2 Illustration of ALARP Process

Risk assessments are intended to ensure that solutions are found in accordance with authority requirements and expectations, internal company requirements and accepted industry practice. It is required that the following aspects are addressed:

1. Are all authority requirements satisfied?
2. Are all internal requirements met?
3. Is the analysed risk level on a par with that of comparable concepts/-solutions?
4. If some requirements or practices are not met, can it be demonstrated that the concept nevertheless does not represent an increased risk level?
5. If quantitative targets are defined, are these met with sufficient margin, in order to enable any possible later increase in analysed risk levels, without the need for extensive changes?
6. Is Best Available Technology (BAT) used?
7. Have solutions been chosen with inherent safety standards?
8. Are there any unsolved problems or areas of concern with respect to risk to personnel and/or working environment, or areas where these two aspects are in conflict?
9. Are there any unsolved problems in relation to serious environmental spill?
10. Is the concept robust with respect to safety?
11. Have aspects of the most recent R&D results and other new experience been considered?

The following considerations should as a minimum be made, and subsequent solutions and actions be implemented:

1. Identify possible technical and/or operational improvements of the installation that may contribute to reduced risk to personnel
   a. without substantial capital or operational costs, or other operational drawbacks
   b. and simultaneously improves operation or maintenance, so that any increase in capital costs is offset by savings in operational costs.

2. Identify possible technical and/or operational improvements that may reduce environmental spill risk
   a. without substantial capital or operational costs, or other operational drawbacks
   b. and simultaneously improves operation or maintenance, so that any increase in capital costs is offset by savings in operational costs.

These evaluations should be made without any cost/ benefit comparison or similar considerations.

When all measures that may fulfil the above criteria have been exhausted, the following evaluations should be performed:

3. Identify possible technical and/or operational improvements that may reduce risk to personnel or environment, but which entails substantially increased capital or operational costs or other operational drawbacks. The following assessments should be made for these alternatives:
   a. Overall expected net present value of all costs and income per statistical fatality averted.
   b. Cost distribution (material damage and delayed/deferred production income) for relevant years, given the occurrence of a major accident, with respect to scenarios that are influenced by the measures being considered.
   c. Overall expected net present value of all costs and income per statistically expected reduced 1000 tons of oil spill.
   d. Cost distribution (clean up costs, compensation claims, *etc.*) for relevant years, given the occurrence of a major oil spill, with respect to scenarios that are influenced by the measures being considered.
   e. Loss of reputation for relevant years, for relevant years, given the occurrence of a major accident or major oil spill, with respect to scenarios that are influenced by the measures being considered.

The values that are computed in Step 3 above may be compared to reference values, if stated, for:

- Cost per statistical fatality averted
- Cost per statistically expected 1000 tons of reduced oil spill
- Maximum loss that the company is able to survive in one single year.

Finally, it should be considered if higher limits may be accepted under special circumstances:

- Higher costs per averted statistical life lost, if the initial risk level is high
- Higher costs per statistically expected 1000 tons of oil spill, if the initial environmental risk is high
- Higher costs per statistically expected 1000 tons of oil spill, if the areas that may be exposed to spills are particularly sensitive.

The present ALARP evaluation is limited to protection of escape ways in case of fire, and is thus limited to personnel safety. Environmental risk is therefore not addressed at all.

B.2.3 Cost-Benefit Analysis

Section 5.4 presented the process and generation of alternatives. Table 5.4 presented the results of the Cost-Benefit Analysis.

B.3 Risk Results

B.3.1 Risk to Personnel

The following are the main risk values for the installation:

A. PLL: 0.0147 fatalities per year
B. FAR: 4.2 per 10^8 manhours
C. Impairment frequency for escape ways: $3.8 \cdot 10^{-4}$ per year

The risk level for personnel may be further illustrated by considering the sources on the installation from which the risk occurs, as well as the mechanisms that may lead to fatalities. Figure B.1 presents the sources in terms of deck levels where the accidents are initiated. Figure B.2 presents the risk contributions in terms of type of accident scenario and deck level where the accident starts.

Figure B.3 presents an *f–N* diagram for the installation, together with an 'isorisk' curve. The shape of the *f–N* curve shows that a relatively high proportion of the risk is due to accidents with high number of fatalities (curve is higher above isorisk line around N=10, compared to around N=1). This implies that there is a higher than usual contribution from accidents with 10 or more fatalities.

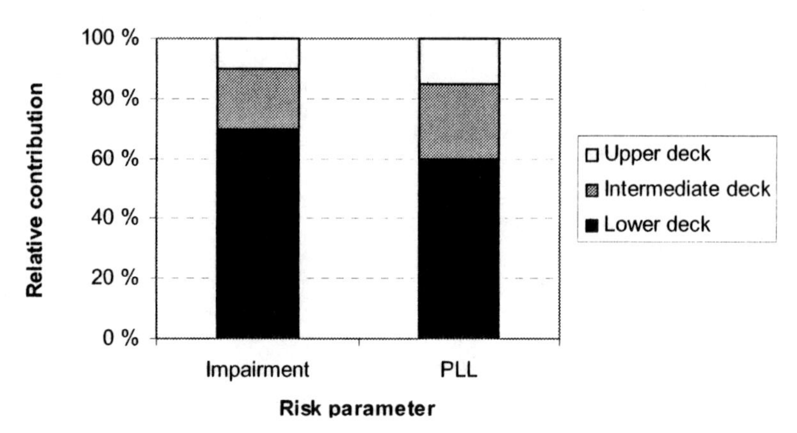

Figure B.1. Contributions to risk from deck levels

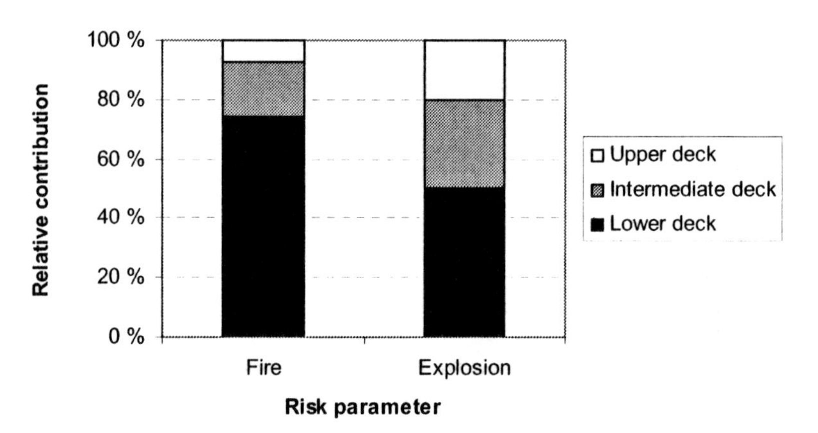

Figure B.2. Contributions to risk from accident scenario and deck levels

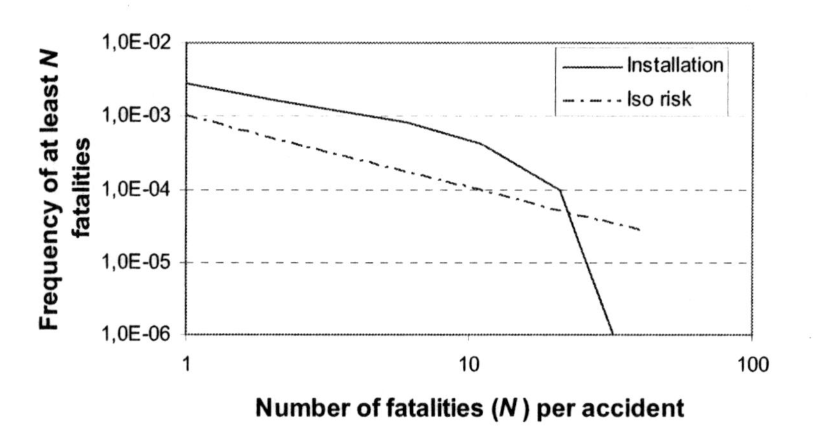

Figure B.3. *f–N* diagram for the case study installation

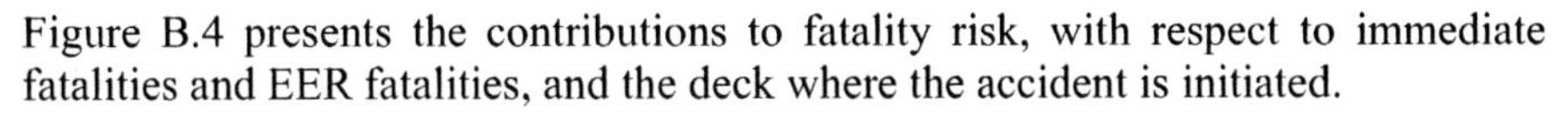

Figure B.4 presents the contributions to fatality risk, with respect to immediate fatalities and EER fatalities, and the deck where the accident is initiated.

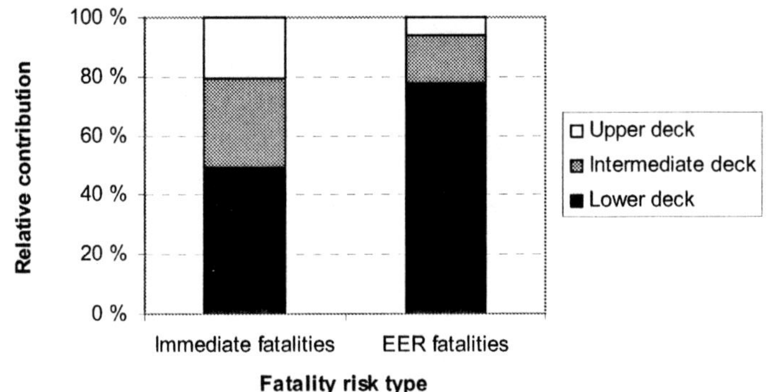

Figure B.4. Type of fatality risk contribution

One of the observations that may be made from the presentations in this section is that the main source of the risk is on the lower deck. The contributions from fatalities during escape, evacuation and rescue (EER) is quite high (40%), and is dominated by failure to escape from the installation.

B.3.2 Risk to Assets

Risk to assets has not been analysed.

B.4 Identification of Possible Risk Reducing Measures

Section B.2.1 presented the sources for qualitative evaluation of risk reducing proposals. A brief summary is given below. These approaches are usually also quite useful during the process of searching for possible risk reducing measures.

It could be observed that there is no formal process for searching extensively for possible risk reducing measures.

B.4.1 Good Practice

Good practice is a general term for good engineering and procedural practices for common situations. It may include solutions, which have not been incorporated into design standards but have been found to be successful in the field. Formal and informal benchmarking is a source of good practice.

B.4.2 Codes and Standards

Codes and standards embody the lessons learnt over past years. They often provide an appropriate solution for well understood hazards and situations. Design codes for particular types of plant and particular services will specify appropriate risk controls and recovery measures.

B.4.3 Engineering Judgement

Engineering judgement involves sound application of engineering and scientific principles and methods to a control situation. It includes within it a subjective experience-based 'feel' for what is acceptable. It is particularly useful for filtering out extremes – situations that are clearly inappropriate – to allow more rigorous analysis of the less clear situations.

B.4.4 Stakeholder Consultation

Consultation with stakeholders–workforce, particularly those exposed, safety representatives, supervisors, managers, and regulators–is an important part of the ALARP judgement, particularly if the views, concerns and perceptions of any of these groups are not aligned.

B.4.5 Tiered Challenge

A team of operations and specialist staff works together down the hierarchy, identifying all the possible control options in each category. The team then starts with the highest one and challenges why it cannot be applied. If the case is made and it is agreed not to apply a control, the team moves on to the next one down. It carries on down the options list, and eventually identifies the option, which is most acceptable to everyone. This is a simplistic description for the sake of illustration. In reality, a combination of controls will normally be required to achieve ALARP. While this sounds quite formal, the discussion throws up relevant information such as remaining life of the facility, profitability, shutdown opportunities, possible process changes and other key inputs to the ALARP decision. The range of the team ensures a widely thought out solution.

B.5 Evaluation of Individual Risk Reducing Measures

B.5.1 List of RRMs

The full range of techniques listed in Section B.4 is employed in the search for risk reducing measures. This has resulted in the following list of RRMs (for illustration purposes the list has been reduced to a manageable length):

1. Limited explosion relief area increase
2. Protective shielding on escape ways
3. Additional escape way
4. Reduction of the number of leak sources
5. Installation of freefall lifeboats
6. Removal of all weather cladding on process modules.

B.5.2 Evaluation of RRMs

Limited Explosion Relief Area Increase
The scope of this proposal is a minor improvement of explosion relief areas around the process modules on the installation, in order to compensate for increased risk due to new equipment, but no further reduction. The proposal will increase ventilation rates slightly and reduce peak explosion overpressures somewhat. This is in accordance with good practice as well as codes and standards.

One benefit of the limited extent of this change is that the working environment conditions (chill factor) for the personnel working in the process areas should not be made dramatically worse. The downside is that the improvement in ventilation is very slight.

The engineering assessment of this proposal is that it should be implemented as a compromise between the need to increase ventilation and the physical working environment conditions. But this measure is not sufficient to solve the critical issue of the lack of protection against smoke for personnel escaping from a fire.

Table B.1 presents the risk results for the calculation of changes resulting from an increase in explosion relief areas, presented against the base case risk results.

Table B.1. Risk values for explosion relief increase

Alternative	Annual impairment frequency (escape ways)	FAR	PLL (/yr)	ΔPLL (/yr)
Base case	$3.76 \cdot 10^{-4}$	4.2	0.0147	
Increase of explosion relief areas	$3.75 \cdot 10^{-4}$	4.4	0.0154	- −0.0007

The annual impairment frequency for escape ways for both alternatives should be stated as $3.8 \cdot 10^{-4}$. The frequencies are presented with a higher number of decimals in this and subsequent tables, in order to show the difference.

Protective Shielding on Escape Ways
This option involves the installation of protective shielding on existing escape ways together with overpressure protection in order to avoid smoke ingress into the enclosed escape ways. With respect to good practice, this option certainly fulfils relevant requirements: shielding and overpressure protection of exposed escape ways is a common solution. Typically on FPSOs, this is the common way to protect an escape way running the full length of the vessel, past the process areas.

On the other hand, this topic is not addressed in the present codes and standards, being regarded as a novel aspect, and therefore not yet reflected in standards.

From an engineering point of view, the judgement should be that protective shielding may be a good solution to high heat loads, but it needs to be combined with overpressure protection in order to ensure that the escape ways are not made unusable due to smoke ingress.

This proposal has been discussed with workforce representatives who support this proposal fully but have questioned whether it is sufficient as a good total solution for safe escape from the installation in the event of serious fire.

The effect on risk results has been calculated, and the results are as shown in Table B.2. It is shown by the results that a substantial reduction in the frequency of impairment of escape ways is due to this improvement, and also a clear reduction in FAR value. The escape ways' impairment frequency is slightly above the limit of 10^{-4} per year (see further description in Section 5.1, page 101), but not significantly above, when uncertainties are taken into consideration.

Table B.2. Risk values for heat shielding on escape ways

Alternative	**Annual impairment frequency (escape ways)**	**FAR**	**PLL (/yr)**	**ΔPLL (/yr)**
Base case	$3.76 \cdot 10^{-4}$	4.2	0.0147	
Protective shielding	$1.25 \cdot 10^{-4}$	3.4	0.0118	0.0029

A workshop in order to discuss possible options with all parties involved has been conducted, with representation from workforce, management, operational management, HES specialists, engineering personnel, *etc.* The consensus in the workshop was that an extra escape way might be the ultimate solution, if heat shielding of escape was not sufficient to reduce the frequency of escape ways impairment. It was noted that this was a borderline issue.

Additional Escape Way

Provision of an additional escape way with sufficient shielding is the ultimate solution to the issue of safe escape in the event of critical fire and/or explosion. It is often the case that the best solution may be found when building a new construction, rather than trying to alter an existing design. A further advantage of the additional escape way is that redundancy is built into the design: it would be unlikely for both escape ways to be impaired at the same time, especially when one of these is thoroughly protected.

It is therefore obvious that this solution satisfies good practice and that engineering judgement gives this proposal a good score. Workforce representatives also favour this proposal, whereas management will probably find it very expensive.

The effect on risk results has been calculated, and the results are as presented in Table B.3. It is shown by the results that a substantial reduction of the frequency of impairment of escape ways is due to this improvement, and also a substantial reduction in FAR value. The escape ways' impairment frequency is slightly below the limit of 10^{-4} per year, but not significantly, when uncertainties are taken into consideration.

Table B.3. Risk values for additional escape way installation

Alternative	**Annual impairment frequency (escape ways)**	**FAR**	**PLL (/yr)**	**ΔPLL (/yr)**
Base case	$3.76 \cdot 10^{-4}$	4.2	0.0147	
Additional escape way	$9.40 \cdot 10^{-5}$	2.5	0.0088	0.0059

The workshop referred to above observed that there is strong opposition to this proposal because of the excessive cost involved in constructing a new escape way.

Reduction of the Number of Leak Sources
The workshop referred to also put forward additional proposals for risk reduction. One of these is a proposal to reduce the number of leak sources. This would entail removing the majority of flanged connections in the process area and replacing them with welded connections. This would be particularly important in the crude oil export pumping area, which is the main source of fire risk with effect on the escape ways.

Removal of leak sources is in accordance with good practice, as well as codes and standards. But an evaluation of this proposal will focus on the side effects due to the fact that this will involve very extensive modification of the process systems. There will be a lot of hot work in the process area, not all of which will be conducted in a shut down and depressurised condition. It is therefore considered that the increase of risk during modifications is substantial, and will probably be considerably higher than the risk reduction per year.

It is further noted that a reduction of the impairment risk by 75% is needed in order to come below the limit of 10^{-4} per year; this means that leak frequencies will also need to be reduced by 75%. It is further noted that export pumps are important as leak sources, and that the relevant leak scenarios are almost impossible to eliminate.

The conclusion from the engineering judgement of this proposal is that it is unlikely to provide the necessary risk reduction and will increase risk substantially during implementation of modifications. This proposal is therefore not recommended for further consideration.

Installation of Freefall Lifeboats
Installation of freefall lifeboats (two boats for redundancy purposes) may reduce the need to use escape ways in case of fire. This is not usually a preferred solution, but may act as a compensatory measure if no other solution can be found. Obviously, the protection of escape ways is unchanged, but the need for escape to the shelter area will be formalistically reduced. On the other hand, this is more a formal solution than a practical, operational one. For escape purposes, escape over bridges to a shelter area on a separate installation is distinctly preferable to using lifeboats, including freefall lifeboats.

Consultation with stakeholders is likely to result in strong opposition from the workforce, and also some reluctance from operational management.

Table B.4 shows the results for the option to install freefall lifeboats. It is shown that the reduction of impairment frequency is substantial, whereas the reduction in FAR (and PLL) is more limited. The reduction is not sufficient in relation to the limit of 10^{-4} per year.

Table B.4. Risk values for installation of freefall lifeboats

Alternative	Annual impairment frequency (escape ways)	FAR	PLL (/yr)	ΔPLL (/yr)
Base case	$3.76 \cdot 10^{-4}$	4.2	0.0147	
Install freefall lifeboats	$1.71 \cdot 10^{-4}$	3.7	0.0130	0.0018

It should further be noted that installation of two freefall lifeboats will be very expensive, probably the most expensive of all the proposals considered.

Remove all Weather Cladding on Process Modules

The removal of all weather cladding on process modules will increase natural ventilation in the process areas. The engineering evaluation of this proposal has concluded that it will have the following effects:

- Reduce ignition probability through reducing gas concentrations and extent of gas cloud within flammable region.
- Reduce explosion overpressure in case of delayed ignition, as a result of having more explosion relief areas.
- Possibly reduce smoke production slightly in the event of fire, due to improved ventilation.
- Working environment in the process area will suffer deterioration due to an increase in wind chill factors.

The increased wind chill factor in the process area is a severe negative factor, which is likely to cause strong opposition from the workforce. In this regard, the proposal is not in line with working environment standards. But improvement of ventilation rates is in line with technical safety principles designed to reduce the effects of hydrocarbon leaks. It is quite common that a compromise has to be found between these two opposing objectives.

Table B.5 presents the revised results for the option to remove all weather cladding. It is noted that some reductions are shown, but the reduction in impairment frequency for escape ways is not at all sufficient to reduce the value below the limit of 10^{-4} per year. This, together with the working environment aspect, means that this proposal will not receive much support from some of the stakeholders.

Table B.5. Risk Values for Removal of all Weather Cladding in Process Area

Alternative	Annual impairment frequency (escape ways)	FAR	PLL (/yr)	ΔPLL (/yr)
Base case	$3.76 \cdot 10^{-4}$	4.2	0.0147	
Remove all weather cladding	$2.37 \cdot 10^{-4}$	3.4	0.0118	0.0029

B.6 Overall Evaluation of Risk Reduction Measures

As a basis for overall decision-making, the following dimensions (see Section 3.3.3, page 83) are taken into account:

A. aspects related to consequences
B. aspects related to uncertainties
C. aspects related to manageability.

The point is that the above calculations express conditional probabilities and expected values $P(A|K)$ and $E[X|K)$, for some events A and unknown quantities X. A may express the occurrence of an accidental event and X may express the number of fatalities next year), given the background information and knowledge K. What we are concerned about are A and X, the actual observable quantities, but our analysis provides only some assignments P and E, which express the analyst's judgements based on K, and could deviate strongly from the observables. Key factors that could lead to such deviations need to be addressed and communicated to management, as a part of the overall description of the risk picture. Sensitivity and robustness analysis are tools that can be used to illustrate the dependence of these factors and the background information K. Some examples of such sensitivity and robustness analyses are presented and discussed below. The main aspects related to the Categories A–C are:

- Given possible fire scenarios; what are the smoke and radiation impacts? What barriers can reduce the possible consequences and avoid fatalities? How reliable and robust are these barriers? Vulnerabilities?

  With the oil export pumps being the main threat, the smoke production from fires will be very dense and poisonous. The heat loads may be limited due to the smoke, but still at such levels that personnel will be fatally injured after only a few seconds.

  The existing escape ways (external vertical towers and external gangways) do not provide any protection of personnel, so that if a fire occurs there are no barriers to protect personnel.
- The analysis assigns a probability of a fire of 1% during a 40 year period. However, a fire may occur, and the additional fire protection will have a considerable positive effect in protecting personnel.

  Even though the frequency of critical fires is as low as 1% over a period of 40 years, the protection of escape ways will also help in less critical fires, which will be somewhat more likely to occur. In the space of 40 years, limited fires may have a probability of typically 50%.
- The company may implement uncertainty and safety management activities that contribute to avoiding the occurrence of hazardous situations and thus accident events. Although there is a risk (expressed by the P and E), diligent efforts are made to avoid events A and obtain desirable outcomes X. These activities are mainly related to human and organisational factors, as well as the HES culture.

  One could argue that most hydrocarbon leaks are due to manual intervention on process equipment. In theory, all non-essential personnel could be removed from all areas where effects could be experienced during the use of escape ways in a fire scenario. Management, however, may consider that this places too much restriction on the operation of the installations, so that this is not feasible in practice.

  On the issue of robustness, it should be noted that heat and smoke protection on escape ways is a passive way of protecting personnel, which does not require any mobilisation or action in an emergency. Therefore, it is usually considered to be a robust way of reducing risk, as opposed to actions that rely on equipment to be started or management actions to be implemented and followed up, which will often have much higher failure probability.

A sensitivity study is a natural part of a broad decision-making process. Some sensitivity study results are presented in Table B.6 and Table B.7.

Table B.6. Results of sensitivity study for heat shielding on existing escape way

Variation	**Resulting Cost/E(life)** (mill NOK)
Base case	315
10 times higher failure frequency for severe fire	32
2 times higher radiation level on escape ways	52
Increased (2 times) proportion of south-westerly wind direction	21
Reduced (50%) proportion of south-westerly wind direction	511

Table B.7. Results of sensitivity study for additional escape way

Variation	**Resulting Cost/E(life)** (mill NOK)
Base case	473
10 times higher failure frequency for severe fire	47
2 times higher radiation level on escape ways	62
Increased (2 times) proportion of south-westerly wind direction	31
Reduced (50%) proportion of south-westerly wind direction	719

The sensitivity study results show considerable variations; some of the results are at such levels that gross disproportion is not an obvious conclusion. This suggests that the analysis is quite sensitive to assumptions and simplifications made in the analysis of risk to personnel.

Many companies have formulated 'zero vision' objectives for their HES management, implying that the long-term objective is to carry out all operations without losses and damage. Sometimes it may be difficult to see the connection between such objectives and the traditional approach to decision-making, involving a narrowly-focused decision-making process with short-term cost minimisation as the driving force.

The decision-making process should enable a broad assessment of potential consequences and uncertainty, so that all the main aspects relating to outcomes of the decisions are available to the decision-makers. The difficult management decision to be taken may be illustrated as follows:

If the decision to install extra protection is taken (with cost of about 36 million NOK), the outcome over the long residual production period (30–40 years) is either one of the following three outcomes:

(a) No fire occurs at all (about 50% probability), and the protection is wasted in terms of pay-back.
(b) A limited fire (not critical fire) occurs (about 49% probability), and the protection has some advantage, thus avoiding any injuries to personnel due to fire loads on escape ways.
(c) A critical fire occurs (about 1% probability), and the protection is very valuable in terms of allowing all personnel to escape to a safe location.

Obviously, if no extra protection is installed, the scenario alternatives are the same, but the outcomes in terms of pay-back are the opposite:

(a) No fire occurs (about 50% probability), no cost, no other effect.
(b) A limited fire (not critical fire) occurs (about 49% probability), the lack of protection means that some of the personnel will be injured during escape, but not fatally.
(c) A critical fire occurs (about 1% probability), the lack of protection implies that more than 50 persons are prevented from escaping to a safe location, many of whom may perish.

If considered in standard economic terms only, the difficult management decision is to consider the 1% probability over a 40 year field lifetime of a severe fire occurring, with possibly up to 30 fatalities, and whether to invest about 36 million NOK in protective systems and actions to avoid these severe consequences.

An argument against the alternative approach to avoid higher investments in risk reduction cannot be accepted if the companies are serious when they formulate 'zero vision' objectives. If a 'zero vision' objective is accepted, it must inevitably be expected that extra costs will be incurred as a consequence. Otherwise the objective should be reworded to 'zero vision as long as it doesn't cost anything'.

B.7 Final Selection of Risk Reduction Measures

The full list of RRMs was presented in Section B.4 as the following:

1. Limited explosion relief area increase
2. Protective shielding on escape ways
3. Additional escape way
4. Reduction of the number of leak sources
5. Installation of freefall lifeboats
6. Removal of all weather cladding on process modules.

It has been concluded through qualitative evaluation that RRM 2 and 3 are alternatives which may achieve the required improvement of safety on the installation. It has further been concluded that RRM 1 and 5 are feasible but insufficient, whereas RRM 4 and 6 are not recommended, because of severe negative effects.

The question is therefore which of the RRMs 1, 2, 3 and 5 that should be decided for implementation. A cost-benefit/cost-effectiveness analysis may be helpful in providing further insight. The cost-benefit/cost-effectiveness analysis is carried out for RRMs 1–3, whereas it is concluded without detailed analysis that the cost of installing freefall lifeboats is grossly disproportionate in relation to the benefits.

Table B.8. Overview of expected cost values for the decision alternatives

Options		Investment cost (million NOK)	Annual operating cost (million NOK)
0	Base case	0	0
1	Limited explosion relief increase	2	0.05
2	Protective shielding	30	0.4
3	Additional escape way	110	0.1

Table B.9 shows that Options 2 and 3 have considerable cost levels per averted statistical life lost (ICAF). If these values are considered in isolation in a quantitative context, the value for Option 3 would usually be considered to be grossly disproportionate in relation to the benefits, the reduction of PLL over 40 years, whereas the value for Option 2 is on the borderline.

Table B.9. Overview of key risk and cost values for the decision alternatives

Options		E[NPV] (40 yrs) (million NOK)	ΔPLL (40 yrs)	Cost/E(life) (million NOK)
0	Base case			
1	Limited explosion relief increase	2.7	0.0	NA
2	Protective shielding	35.7	0.1	303
3	Additional escape way	111.4	0.2	473

Figure B.5 compares all risk reduction alternatives with respect to risk reduction, expressed as FAR and impairment frequency for escape ways.

B.8 Risk Levels after Implementation of Measures

B.8.1 Measures Accepted for Implementation

Based on the evaluations and assessments reported above, the following measures have been proposed for implementation:

1. Limited explosion relief area increase
2. Protective shielding on escape ways.

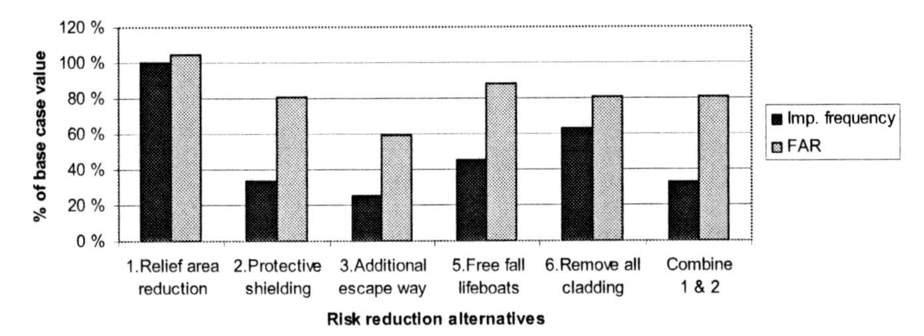

Figure B.5. Comparison of risk reduction alternatives with respect to personnel risk

Item 1 is included because it makes good sense, and has a limited cost. Item 2 is chosen in preference to Item 3 because it is less expensive, has a somewhat lower ICAF value, and brings the impairment frequency for escape ways down to the order of magnitude of 10^{-4} per year.

This proposal has been made by management and discussed with all stakeholders, who have unanimously agreed that the proposal is a reasonable compromise.

B.8.2 Measures Not Accepted for Implementation

The measures not accepted for implementation should be documented explicitly, as noted in Section B.2.1. The following is a summary of the arguments against accepting the RRMs 3, 4, 5 and 6 in the list shown in Section B.7:

3.	Additional escape way	This proposal is considerably more expensive than RRM2 (protective shielding). It also has a lower ICAF value than RRM2, and is therefore less efficient.
4.	Reduction of the number of leak sources	This proposal is unlikely to provide as much risk reduction as needed. There is also a significant increase of risk during the implementation phase, due to extensive use of hot work.
5.	Installation of freefall lifeboats	This proposal does not solve the actual problem; it provides a more formalistic solution. It is in addition very expensive.
6.	Remove all weather cladding on process modules	The proposal is not in line with working environment standards, and is not likely to reduce the smoke problem sufficiently.

B.8.3 Residual Risk for Personnel

Table B.10 presents the resulting risk levels for personnel, which reflects the proposed risk reduction actions as listed above.

Table B.10. Risk values for heat shielding on escape ways

Alternative	Annual impairment frequency (escape ways)	FAR	PLL (/yr)	ΔPLL (/yr)
Base case	$3.76 \cdot 10^{-4}$	4.2	0.0147	
Protective shielding and relief areas increased	$1.22 \cdot 10^{-4}$	3.4	0.0120	0.0027

B.8.4 Overall Evaluation of Risk

The overall evaluation of risk is focused on the aspects presented in Section B.2.2. The following evaluations may be performed:

1. *Are all authority requirements satisfied?*
   With the proposed actions, the impairment frequency is reduced to a level on a par with existing regulations, which implies that the protection of escape ways corresponds to that of modern installations.
2. *Are all internal requirements met?*
   All internal requirements by the operator have been met.
3. *Is the analysed risk level on a par with that of comparable concepts/ solutions?*
   The risk level is presented above, and is low in general compared with other large process installations. The FAR value is low, which reflects the fact that the installation is part of a 'production complex', with bridge connection to a separate installation for accommodation purposes.
4. *If some requirements or practices are not met, can it be demonstrated that the concept nevertheless does not have an increased risk level?*
   All requirements are satisfied.
5. *If quantitative targets are defined, are these met with sufficient margin, in order to allow for any later increase in analysed risk levels, without the need for extensive changes?*
   It may be argued that there is no margin with respect to the limit for impairment of escape ways. However, the installation is not in the design phase, which makes future extensions unlikely.
6. *Is Best Available Technology (BAT) used?*
   Best Available Technology would entail separating personnel completely from fire exposure. This is certainly not possible in the case of an existing installation and often not completely possible on a new installation either.
7. *Have solutions been chosen with inherent safety standards?*
   Inherent safety would be better implemented through provision of a new escape way; however, it is considered that this is too expensive on an existing installation.
8. *Are there any unsolved problems or areas of concern with respect to risk to personnel and/or working environment, or areas where these two aspects are in conflict?*

There is no aspect which is considered unsolved with respect to personnel risk and/or working environment. If complete removal of all weather cladding had been chosen, this would have been a conflict area.

9. *Are there any unsolved problems in relation to serious environmental spill?*
   Not relevant in the present case.
10. *Is the concept robust with respect to safety?*
    A new escape way would have been a more robust solution, but the heat shielding is considered to be a reasonable compromise for an existing installation.
11. *Are aspects of the most recent R&D results and other new experience considered?*
    Not relevant in the present case.

B.9 Implementation Plan for Measures

The final step in the ALARP demonstration is the implementation plan for the agreed risk reduction measures, as suggested in Table B.11.

Table B.11. Implementation plan for agreed risk reduction measures

I.D. No.	Description	Studied by	Imple-mented by
1	Limited explosion relief increase in process area	01.10.200X	31.12.200X
2	Protective shielding on escape ways	31.12.200X	01.06.200Y

References

Abrahamsen, E.B., Aven, T., Vinnem, J.E. and Wiencke, H.S. (2004) Safety Management and the use of expected values. Risk, Decision and Policy 9:347–358.–

Abrahamsen, E.B., Asche, F. and Aven, T. (2005) A discussion of the principles of cost-benefit analyses for analysing safety measures. In: Kolowrochi, K. (ed.). Proceedings ESREL 2005. Balkema, London, pp. 15–20.

Abrahamsen, E.B. and Aven, T. (2006a) On the use of cost-benefit analysis and the cautionary principle in safety management. In: Stamatelates, M.G. and Blackman, H.S. (eds). Proceedings PSAM 8. ASME Press, New York.

Abrahamsen, E.B. and Aven, T. (2006b) On the consistency of risk acceptance criteria with normative theories for decision-making. In: Soares, C. (ed.). Proceedings ESREL 2006. Balkema, London.

Allais, M. and Hagen, O. (eds) (1979) Expected Utility Hypothesis and the Allais Paradox. D. Reidel, Dordrecht.

Apostolakis, G. (1990) The concept of probability in safety assessments of technological systems. Science 250:1359–1364.

Apostolakis, G., Farmer, F.R. and van Otterloo, R.W. (eds) (1988) The interpretation of probability in probability safety assessments. Reliability Engineering & System Safety 23, Issue 4.

Apostolakis, G. and Wu, J.S. (1993) The interpretation of probability. De Finetti's representation theorem, and their implication to use of expert opinions in safety assessment. In: Reliability and Decision-making, Barlow, R.E. and Clarooti, C.A. (eds). Chapman & Hall, London, pp. 311–322.

Asche, F. and Aven, T. (2003) On the economic value of safety. Risk Decision and Policy 9:253–267.

Aven, T. (1994) On safety management in the petroleum activities on the Norwegian Continental Shelf. Reliability Engineering & System Safety 45:285–291.

Aven, T. (2003) Foundations of Risk Analysis: A Knowledge and Decision-Oriented Perspective. Wiley, New York.

Aven, T. (2006a) On the precautionary principle, in the context of different perspectives on risk. Risk Management: An International Journal 8:192–205.

Aven, T. (2006b) On the ethical justification of the use of risk acceptance criteria. Risk Analysis, to appear.

Aven, T. and Abrahamsen, E. (2006) On the use of cost-benefit analysis in ALARP processes. Paper submitted for publication.

Aven, T., Hauge, S. Sklet, S. and Vinnem, J.E. (2006a) Methodology for incorporating human and organisational factors in risk analyses for offshore installations, International Journal of Materials & Structural Reliability 4:1–14.

Aven, T. and Kørte, J. (2003) On the use of cost/benefit analyses and expected utility theory to support decision-making. Reliability Engineering & System Safety 79:289–299.

Aven, T. and Kristensen, V. (2005) Perspectives on risk – review and discussion of the basis for establishing a unified and holistic approach. Reliability Engineering & System Safety 90:1–14.

Aven, T., Nilsen, E. and Nilsen, T. (2004) Economic risk – review and presentation of a unifying approach. Risk Analysis 24:989–1006.

Aven T. and Pitblado R. (1998) On risk assessment in the petroleum activities on the Norwegian and UK continental shelves. Reliability Engineering & System Safety 61:21–29.

Aven, T. and Vinnem, J.E. (2005) On the use of risk acceptance criteria in the offshore oil and gas industry. Reliability Engineering & System Safety 90:15–24.

Aven, T., Vinnem, J.E. and Røed, W. (2006b) On the use of goals, quantitative criteria and requirements in safety management. Risk Management: An International Journal 8:118–132.

Aven, T., Vinnem, J.E. and Vollen, F. (2006c) Perspectives on risk acceptance criteria and management for offshore installations – application to a development project. International Journal of Materials & Structural Reliability 4:15–25.

Aven, T., Vinnem, J.E. and Wiencke, H.S. (2006d) A decision framework for risk management. Reliability Engineering and System Safety, to appear.

Barlow, R.E. (1998) Engineering Reliability. SIAM, Philadelphia.

Bedford, T. and Cooke, R. (2001) Probabilistic Risk Analysis: Foundations and Methods. Cambridge University Press, Cambridge.

Bernardo, J.M. and Smith, A. (1994) Bayesian Theory. Wiley, New York.

Beck, U. (1992) Risk Society. SAGE Publications, London.

Cabinet Office (2002) Risk: Improving Government's Capability to Handle Risk and Uncertainty. Strategy Unit Report, UK.

Chapman, C. and Ward, S. (1997) Project Risk Management: Processes, Techniques and Insights. Wiley, Chichester.

Chapman, C. (1997) Project risk analysis and management: PRAM the generic process. International Journal of Project Management 15:273–281.

Cherry, J. and Fraedrich, J. (2002) Perceived risk, moral philosophy and marketing ethics: Mediating influences on sales managers' ethical decision-making. Journal of Business Research 55:951–962.

Clemen, R.T. (1996) Making Hard Decisions, 2nd edition. Duxbury Press, New York.

Cooke, R.M. (1991) Experts in uncertainty: Opinion and Subjective Probability in Science. Oxford University Press, New York.

Cox, L.A. (2002) Risk Analysis: Foundations, Models and Methods. Kluwer, Boston.

Douglas, M. and Wildavsky, A. (1982) Risk and Culture. University of California Press, Berkeley.

Douglas, E.J. (1983) Managerial Economics: Theory, Practice and Problems, 2nd edition. Prentice Hall, Englewood Cliffs NJ.

EAI (2006) Risk and uncertainty in cost benefit analysis. A toolbox paper for the Environmental Assessment Institute. http://www.imv.dk.

EU (European Union) Commission (2000) Communication from the Commission on the Precautionary Principle. Brussels February 2, 34, Com(2000)1.

Evans, A.W. and Verlander, N.Q. (1997) What is wrong with criterion FN-lines for judging the tolerability of risk? Risk Analysis 17:157–168.

Fischhoff, B., Lichtenstein, S., Slovic, P., Derby, S. and Keeney R. (1981) Acceptable Risk. Cambridge University Press, New York.

Gollier, C., Jullien, B. and Treich, N. (2000) Scientific progress and irreversibility: An economic interpretation of the Precautionary Principle, Journal of Public Economics 75:229–253.

Gray, J.S. (1998) Statistics and the Precautionary Principle. Marine Pollution Bulletin 21:174–176.

Greenpeace (1996) Brent spar protest in the North Sea, 22.5.1996 http://archive.greenpeace.org/comms/brent/brent.html.

Hattis, D. and Minkowitz, W.S. (1996) Risk evaluation: criteria arising from legal traditions and experience with quantitative risk assessment in the United States. Environmental Toxicology and Pharmacology 2:103–109.

Hauge, S., Hokstad P. and Onshus T. (2001) The introduction of IEC61511 in Norwegian offshore industry. In: Zio, E., Remichela, M. and Piccinini, N. (eds). ESREL 2001:483–490.

Henley, E.J. and Kumamoto, H. (1981) Reliability Engineering and Risk Assessment. Prentice-Hall, New York.

Hill, R.H. Jr. (2003) The safety ethic: Where can you get one? Chemical Health and Safety 10:8–11.

Hokstad, P., Vatn, J., Aven, T. and Sørum, M. (2004) Use of risk acceptance criteria in Norwegian offshore industry, Dilemmas and challenges. Risk, Decision and Policy 9:193–206.

Hokstad, P. and Steiro, T. (2005) Overall strategy for risk evaluation and priority setting of risk regulations. Reliability Engineering & System Safety 91:100–111.

Hovden (1999) Ethics and safety: Mortal questions for safety management. Safety Science Monitor 3:1–9.

HSE (1992) Offshore Installations (Safety Case) Regulations, 1992. HSE books, London.

HSE (2001a) Reducing risks, protecting people, HSE's decision-making process [R2P2]. HSE books, London. http://www.hse.gov.uk/dst/r2p2.pdf

HSE (2001b) Principles and guidelines to assist HSE in its judgements that duty-holders have reduced risk as low as reasonably practicable. HSE books, London. www.hse.gov.uk/risk/theory/alarp1.htm

HSE (2002) HID's approach to "As Low As Reasonably Practicable" (ALARP) decisions, SPC/Permissioning/09. HSE books, London. http://www.hse.gov.uk/comah/circular/perm09.htm

HSE (2003a) Assessing compliance with the law in individual cases and the use of good practice. HSE books, London. www.hse.gov.uk/risk/theory/alarp2.htm

HSE (2003b) Policy and guidance on reducing risks as low as reasonably practicable in design. HSE books, London. www.hse.gov.uk/risk/theory/alarp3.htm

HSE (2003c) Guidance on ALARP for offshore division inspectors making an ALARP demonstration, SPC/Enf/38. HSE books, London. http://www.hse.gov.uk/offshore/circulars/enf38.htm

HSE (2006) Offshore Installations (Safety Case) Regulations, 2005. HSE books, London.

IEC 61508 (2000) Functional safety of electrical/electronic/programmable electronic (E/E/PE) safety related systems, parts 1–7. International Electrotechnical Commission, Geneva, Switzerland.

IEC 61511 (2003) Functional safety: Safety instrumented system for the process industry sector, part 1–3, December 2003. International Electrotechnical Commission, Geneva, Switzerland.

ISO (2002) Risk management vocabulary. ISO/IEC Guide 73.

ISO (1999) Petroleum and natural gas industries – Collection and exchange of reliability and maintenance data for equipment. ISO 14224:1999.

Kaplan, S. (1992) Formalism for handling phenomenological uncertainties: The concepts of probability, frequency, variability, and probability of frequency. Nuclear Technology 102:137–142.

Kaplan, S. and Garrick, B.J. (1981) On the quantitative definition of risk. Risk Analysis 1:1–27.

Kasperson, R.E. (1992) The social amplification of risk: process in developing an interrative framework. In Krimsky, S. and Golding D. (eds). Social Theories of Risk. Praeger, Westport pp. 153–178.

Kastenberg, W., Hauser-Kastenberg, G. and Norris, D. (2004) On developing a risk analysis framework for post-industrial age technologies. In: Spitzer, C., Schmocker, U. and Dang, V.N. (eds). PSAM7. Springer, London, pp. 2378–2383.

Klinke, A. and Renn, O. (2001) Precautionary principle and discursive strategies: classifying and managing risk. Journal of Risk Research 4:159–173.

Kristensen, V., Aven, T. And Ford, D. (2006) A new perspective on Renn & Klinke's approach to risk evaluation and risk management. Reliability Engineering & System Safety 91:421–432.

Kvaløy, J.T. and Aven, T. (2005) An alternative approach to trend analysis of accident data. Reliability Engineering & System Safety 90:75–82.

Levy, H. and Sarnat, M. (1990) Capital Investment and Financial Decisions. Fourth edition. Prentice Hall, New York.

Lind, N. (2002) Social and economic criteria of acceptable risk. Reliability Engineering & System Safety 78:21–26.

Lindley, D.V. (1985) Making Decisions. Wiley, New York.

Lindley, D.V. (2000) The philosophy of statistics. The Statistician 49:293–337.

Lofstedt, R. E. (2003) The Precautionary Principle: risk, regulation and politics. Transactions IchemE 81:36–43.

Modarres, M. (1993) What Every Engineer Should Know about Reliability and Risk Analysis. Marcel Dekker, New York.

Morris, J. (ed.) (2000) Rethinking Risk and the Precautionary Principle. Butterworth-Heinemann, Oxford.

Melchers, R.E. (2001) On the ALARP approach to risk management. Reliability Engineering & System Safety 71:201–208.

Nilsen, T. And Aven, T. (2003) Models and model uncertainty in the context of risk analysis. Reliability Engineering & System Safety 79:309–317.

NORSOK (2001) Risk and Emergency Preparedness Analysis. NORSOK Standard Z–013.

OHSAS 18002:2000. Occupatonal health and safety management systems – Guidelines for the implementation of OHSAS 18001. ISBN:0580331237.

Okrent, D. and Pidgeon, N. (1998) Special issue on Risk perception versus risk analysis. Reliability Engineering & System Safety 59.

OLF (2001) OLF guideline 070 on the application of IEC 61508 and IEC 61511 in the petroleum activities on the Norwegian Continental Shelf, OLF, Rev 01; http://www.itk.ntnu.no/sil

OSPAR (1992) The convention for the protection of the marine environment of the North-East Atlantic. Paris, 22 September 1992 http://www.ospar.org/eng/html/welcome.html.

Pape, R.P. (1997) Developments in the tolerability of risk and the application of ALARP. Nuclear Energy 36:457–463.

Pearce, D. (1994) The Precautionary Principle in economic analysis. In: Cameron, J. and O'Riordan, T. (eds). Interpreting the Precautionary Principle. Cameron May, London.

Perrow, C. (1984) Normal Accidents. Basic Books, New York.

Pidgeon, N.F. and Beattie, J. (1998) The psychology of risk and uncertainty. In: Calow, P. (ed.). Handbook of Environmental Risk Assessment and Management. Blackwell Science, London, pp. 289–318.

PSA (2000) Risk Level on the Norwegian Continental Shelf, Methodology report 2000, Petroleum Safety Authority Norway 16.5.2000.

PSA (2001a). Regulations relating to Management in the Petroleum Activities (the Management Regulations), Petroleum Safety Authority.

PSA (2001b) Guidelines to Management Regulations, Petroleum Safety Authority.

PSA (2002) Framework Regulations, www.ptil.no.

PSA (2005) Risk Level on the Norwegian Continental Shelf, Main report 2004, Phase 5, SPA report 05–02, 26.4.2005.

Renn, O. (1992) Concepts of risk: a classification. In: Krimsky, S. and Golding, D. (eds). Social Theories of risk. Praeger, Westport, pp. 53–79.

Renn, O. and Klinke, A. (2002) A new approach to risk evaluation and management: Risk-based precaution-based and discourse-based strategies. Risk Analysis 22:1071–1094.

Rimington, J., McQuaid, J. and Trbojevic, V. (2003) Ministerie van Sociale Zaken en Werkgelegenheid, Application of Risk-Based Strategies to Workers' Health and Safety Protection – UK Experience, August 2003, ISBN 90–5901–275–5.

Rodgers, M.D. (2001) Scientific and technological uncertainty, the precautionary principle, scenarios and risk management. Journal of Risk Research 4:1–15.

Rosa, E. A. (1998) Metatheoretical foundations for post-normal risk. Journal of Risk Research 1:15–44.

Sandin, P. (1999) Dimensions of the precautionary principle. Human and Ecological Risk Assessment 5:889–907.

Sandøy, M., Aven, T. and Ford, D. (2005) On integrating risk perspectives in project management. Risk Management: An International Journal 7:7–21.

Schofield, S. (1998) Offshore QRA and the ALARP principle. Reliability Engineering & System Safety 61:31–37.

Shrader-Frechette, K.S. (1991) Risk and Rationality. University of California Press, Berkeley.

Singpurwalla, N.D. (1988) Foundational issues in reliability and risk analysis. Society for Industrial and Applied Mathematics 30:264–281.

SINTEF (1992) Accident costs in the industry (In Norwegian). STF75 A92032.

Skjong, R. and Ronold, K.O. (2002) So much for safety. Proceedings of OMAE. Oslo, 23–28 June, 2002.

Stern, P.C. and Fineberg, H.V. (eds) (1996) Understanding Risk. National Academy Press, Washington DC.

Storting (2003) Investigation of the working conditions for the pioneer divers in the north Sea (in Norwegian only). White paper No 47 (2002–2003), 27.6.2003.

Thomen, J.R. (1991) Leadership in Safety Management. John Wiley, New York.

Thompson, P.A. and Perry, J.G. (1992) Engineering Construction Risks: A Guide to Project Risk Analysis and Risk Management. Thomas Telford, London.

Total (2003) Frigg field cessation plan. Total E&P Norge, 9.5.2003. http://www.total.no/no/Activities/Field+decommissioning/index.aspx

UKOOA (1999) A framework for risk related decision support – Industry guidelines. UK offshore Operators Association.

Varian, H.R. (1999) Intermediate Microeconomics: A Modern Approach. 5th edition. W.W. Norton & Company, New York.

Vatn, J. (1998) A discussion on the acceptable risk problem. Reliability Engineering & System Safety 61:11–19.

Vinnem, J.E. (2000) Risk monitoring for major hazards. SPE61283, SPE International Conference on Health, Safety and the Environment in Oil and Gas Exploration and Production in Stavanger, Norway 26–28 June 2000.

Vinnem, J.E. (2006) On the analysis of operational barriers on offshore petroleum installations. Proceedings of the 8th International Conference of Probabilistic Safety Assessment and Management, New Orleans, Louisiana.

Vinnem, J.E. (2007) Offshore Risk Assessment. 2nd edition. Springer Verlag, London, to appear.

Vinnem, J.E. and Aven, T. (2001): Methodology for evaluation of risk level for societal activities, examples illustrating possible approaches (in Norwegian only). Preventor/RF report 2000048–01, 22.9.2001.

Vinnem J.E., Aven, T. and Wiencke, H.S. (2006a) Case illustration of a decision framework for HES decisions. Journal of Risk and Reliability, in press.

Vinnem, J.E., Aven, T., Husebø, T., Seljelid, J. and Tveit, O. (2006b) Major hazard risk indicators for monitoring of trends in the Norwegian offshore petroleum sector. Reliability Engineering & Systems Safety 91:778–791.

Vinnem, J.E., Kristensen, V. and Witsø, E. (2006c). ALARP processes. In: Soares, C. (ed.). Proceedings ESREL 2006. Balkema, London.

Vinnem, J.E., Pedersen, J.I. and Rosenthal, P. (1996) Efficient risk management: use of computerised QRA modell for safety improvements to an existing installation. SPE – paper 35775. 3rd International Conference on HSE, New Orleans, 9–12 june 1996.

Viscusi, W.K. (1986) Market incentives for safety. Harvard Business Review 63:133–138.

Viscusi, W.K. (1993) The value of risk to life and health. Journal of Economic Literature 31:1912–1946.

Vrijling, J.K., Hengel, W. and van Houben, R.J. (1998) Acceptable risk as a basis for design. Reliability Engineering & System Safety 59:141–150.

Vrijling, J.K. and van Geldeer, P.H. (1997) Societal risk and the concept of risk aversion. In: Soares, C. (ed.). Proceedings ESREL'97. Elsevier, Oxford.

Vose, D. (2000) Risk Analysis: A Quantitative Guide. Wiley, Chichester.

Wakker, P. (1994) Separating marginal utility and probabilitistic risk aversion. Theory and decision 36:1–44.

Watson, S.R. (1994) The meaning of probability in probabilistic safety analysis. Reliability Engineering & System Safety 45:261–269.

Wiener, J.B. and Rogers, M.D. (2002) Comparing precaution in the United States and Europe, Journal of Risk Research 5:317–349.

Yaari, M. (1987) A dual theory of choice under risk, Econometrica 55:95–115.

Øien, K. and Sklet, S. (2001) Risk analyses during operation (The Indicator Project) – Executive Summary. SINTEF Report STF38 A01405, SINTEF Industrial Management, Norway, 2001.

Øien, K. (2001) Risk indicators as a tool for risk control. Reliability Engineering & System Safety 74:129–145.

Index

A
acceptable risk; 23; 50; 52; 61; 63; 67; 68; 79
accountability; 154
ALARP; v; vii; 3; 16; 20; 35; 50; 62; 68; 69; 75; 76; 83; 88; 96; 99; 102; 103; 104; 111; 115; 117; 149; 156
aleatory uncertainty; 39
Alexander L. Kielland; 94
attribute; v; 29; 79; 86; 151
availability; 22; 140; 141; 146

B
background information; 21; 33; 37; 39; 44; 47; 58; 78; 80; 89; 107
ballast control room; 13; 93
Barents Sea; 4; 5; 9; 10; 11; 32; 33; 34; 36; 37; 40; 153
Bayesian approach; 19; 23; 150; 155
Bayesian paradigm; 70
blowout; 10; 12; 36; 48; 126
BORA; 156
buoyancy; 4; 5; 13; 94; 95; 96; 97; 98; 99

C
capacity; 44; 68; 98; 140; 141; 146
CAPM; 28; 55
cash flow; 5; 6; 27; 28; 30; 43; 46; 47; 55; 56
cautionary principle; 19; 34; 35; 37; 41; 55; 72; 96; 109; 152
CDRS; 130
centre of gravity; 20; 31
classical perspective; 73
CODAM; 130
codes and standards; 86; 97; 104; 158
collision; 97; 98; 99; 121; 122; 130; 131; 134; 138; 139
complexity; vi; 87; 119; 151; 153
conditional probability; 133
consequence analysis; 9
consistency; 4; 76; 154
continuous improvement; 2; 146; 154
cost-benefit analysis; 7; 26; 34; 41; 54; 69; 72; 99; 152; 158
cost-effectiveness analysis; 9; 19; 26; 28; 29
CSE; 52

D
decision analysis; 22; 26; 31; 32; 156
decision criteria; 2; 65; 67; 116
decision principles; 82; 83; 89; 113; 115; 154
decision problem; 24; 26; 34; 44; 49; 77; 79; 82; 86; 90; 102; 104; 109; 113; 118; 155
decision process; 4; 41; 57; 65; 76; 81; 154; 156
decision-making; v; 1; 11; 13; 16; 19; 20; 24; 41; 46; 56; 60; 65; 67;

69; 72; 81; 89; 91; 93; 95; 96; 98; 102; 106; 108; 114; 116; 120; 122; 127; 150; 152; 153; 158
decommissioning; vii; 4; 5; 16; 113; 115; 120; 123
delay effect; 88
DFU; 129; 130
discount rate; 4; 28; 30; 47; 55; 79
discourse; 26; 69; 74; 156
diversification; 30; 43
diving operations; 4; 7; 9

E

economic incentives; 7
economic project evaluations; 5; 6; 12
effectiveness; v; 4; 9; 19; 20; 26; 48; 51; 53; 55; 57; 59; 61; 62; 79; 88; 97; 140; 141; 146; 154
engineering judgement; 55; 97; 104; 159
environmental risk; 5; 10; 33; 34; 50; 62; 68; 73
escalation; 143
escape ways; 15; 63; 101
ESD; 143; 144
ethics; 10; 20; 68
ethics of the consequence; 20; 68; 70; 72
ethics of the mind; 20; 68; 69
evacuation; 63
evaluation of risk; 95
event tree analysis; 132
expected utility theory; 19; 24; 26; 40; 55; 75; 152
expected value; v; 5; 13; 19; 22; 23; 27; 40; 78; 80; 85; 99; 107; 150; 151
experience data; 119
expert group; 14; 33; 48; 49; 146
explosion; 15; 101; 104; 105; 130; 131; 140
exposed hours; 53; 101

F

FAR; 15; 53; 61; 95; 97; 101; 106; 133
field development; 4; 5; 12; 111
fire; 15; 16; 31; 34; 35; 45; 68; 101; 107; 126; 129; 132; 140; 142; 145
fire load; 110
flexibility; 26; 35; 60; 61; 88; 94; 119
floating production; 13; 94; 98; 134
flotel; 94
framing; 96
Frigg; 16; 113; 118; 121; 123
functionality; 84

G

goal; v; 1; 15; 50; 52; 60; 67; 73; 80; 102; 125; 128
good practice; 58; 97; 98; 99; 104; 157; 158
government; vi; 4; 5; 6; 10; 23; 32; 36; 82; 83; 114; 122; 154; 156

H

hazardous situation; 14; 31; 34; 35; 69; 108
HCLIP; 130
health problems; 7
HES culture; 88; 108; 128; 151
HES management; 81; 103; 109; 111; 150; 152
HSE; 38; 41; 42; 54; 75; 76; 153; 157
H-value; 125

I

ICAF; 97; 106; 108
ignition; 129; 140; 143
impairment; 15; 20; 52; 62; 73; 101; 105; 109
impairment frequency; 15; 52; 101; 105; 109
indicator; 132; 136; 138; 139; 143; 145
individual risk; 61; 69; 73
initiating events; 131; 134
inspection; 45; 53; 120; 141
insurance; 2; 8; 31; 54; 70; 107
ISO; viii; 1; 2; 17; 20; 142

K
knowledge uncertainty; 39; 40

L
labour organisations; 4; 13
lagging indicator; 140
leading indicator; 140
leak; 15; 32; 45; 101; 131; 136; 143
level of detail; 61

M
maintenance; 35; 45; 57; 79; 83; 104; 141; 142; 143
major accident; viii; 20; 38; 52; 53; 68; 117; 125; 126; 130; 141; 142; 148; 151
major hazard; 96; 126; 127; 132; 135; 148
manageability; v; vi; 6; 13; 46; 48; 49; 79; 80; 86; 88; 90; 107; 118; 122; 151; 152; 153
managerial review and decision; 82; 96
managerial review and judgement; vi; 4; 74; 77; 78; 89; 153
manhours; 125; 133; 135
minimum safety standards; 74
minor damage; 98
minor effect; 47
mobile units; 125; 134; 136
modelling uncertainty; 38
muster drills; 144

N
Net present value; see NPV
NMD; 13; 94; 98
NORSOK; 53; 152
North Sea; 5; 6; 7; 11; 17; 60; 62; 94; 113; 116; 118
Norwegian Continental Shelf; 4; 5; 6; 9; 10; 52; 63; 126; 134; 136; 138; 143; 145; 146; 154
NPD; 52; 125
NPD regulations; 52
NPV; 5; 12; 27; 28; 29; 30; 41; 42; 46; 47; 49; 79; 86; 97; 106; 152

O
observable quantity; 44
Ocean Ranger; 13; 93; 98
offshore activities; 7
OSPAR; 16; 113; 114; 115; 122

P
persistency; vi; 88
personnel risk; 98
Petroleum Safety Authority; see PSA
PLL; 105; 106
Portfolio theory; 19
precautionary principle; v; vii; 5; 7; 11; 19; 35; 36; 37; 72; 73; 75; 83; 150; 154; 156
prediction; 39; 134; 136; 137; 138; 144; 151
prediction interval; 134; 136; 137; 138; 144
probability; 21; 161
problem definition; vi; 77
process area; 140; 143
project phase; v; vii; 6; 19; 111; 123
PSA; 10; 13; 51; 68; 69; 83; 91; 99; 125; 126; 130; 134; 140; 145; 146; 155; 157; 158
public domain; 93; 113
pump room; 13; 93

Q
QRA; 95; 98; 130; 132; 133; 134; 158

R
real risk; 8; 21; 36; 37; 71; 72
recovery time; 36; 37; 40
relative risk indicator; 135
reliability; 67; 140; 141; 142; 146
reputation; 44; 59; 83; 150
rescue; 25
reservoir uncertainty; 5; 12
reversibility; 88
riser failure; 44; 45; 46
risk acceptance criteria; vii; 3; 4; 20; 50; 57; 58; 60; 95; 101; 111; 150
risk analysis; vii; 6; 8; 10; 14; 21; 36; 44; 51; 56; 61; 64; 66; 70; 75;

78; 87; 95; 101; 129; 131; 149; 156
risk assessment; 1; 2; 4; 5; 7; 13; 14; 17; 23; 42; 53; 56; 60; 62; 78; 86; 95; 102; 153; 155
risk assignments; 71; 72; 78
risk aversion; v; vii; 11; 19; 30; 31; 44; 150
risk communication; 23; 156
risk evaluation; 3; 57; 78
risk indicator; vii; 127
risk level; 3; 14; 15; 17; 21; 59; 60; 61; 64; 68; 70; 74; 79; 95; 101; 104; 127; 132; 139; 140; 142
Risk Level Project; 125; 132; 134; 137; 142; 146; 156
risk management; v; vi; vii; viii; 1; 8; 17; 19; 22; 43; 47; 66; 77; 93; 107; 149; 152; 159
risk perception; 8; 14; 15; 23; 59; 73; 127
risk perspectives; 20
risk reducing measures; vii; 3; 11; 13; 50; 52; 53; 63; 81; 103; 105; 110; 152; 156
risk reduction; 16; 23; 51; 53; 61; 66; 79; 81; 99; 102; 106; 110; 115; 158
risk to personnel; 50; 59; 60; 83; 95; 104; 109; 117; 122
risk transfer; 90
risk treatment; 1; 17; 78
robustness; 44; 97; 98; 107; 108; 140; 141; 146; 152

S

Safety Case Regulations; 157
safety culture; vii; 88; 140
safety function; 15; 53; 62; 73; 101
safety management; vi; vii; 3; 5; 15; 19; 20; 34; 46; 47; 63; 73; 79; 80; 88; 108; 151; 152; 156
science; 7; 10; 37; 38; 41; 127; 156
scientific certainty; 19; 35; 41; 72
sensitivity analysis; 42; 153
shuttle tanker; 130; 131
shuttling; 63; 64; 65; 74; 75; 128
significant damage; 62
significant effect; 16
smoke; 16; 101; 105; 107; 111; 142
social construction; 23
society; 1; 7; 11; 26; 74; 79; 115; 152; 154; 156
SSIV; 44; 45; 46
stakeholder; 17; 83; 90; 97; 104; 115; 122; 155; 156; 159
stakeholder consultation; 97; 104; 122; 159
stochastic uncertainty; 39
strategies; 82; 83; 89; 115; 122; 149; 154
structural failure; 5; 94
supply vessel; 98
systematic risk; 5; 6; 29; 30; 42; 55

T

tiered challenge; 97; 104; 159
transparency; 154

U

ubiquity; vi; 88
UK Health and Safety Executive; 117; 153
uncertainties; 21; 37; 151; 161
uncertainty management; 3; 6; 13; 19; 46; 79; 80; 82
unsystematic risk; 5; 6; 12; 19; 29; 30; 43; 46; 47; 48; 55
utilitarianism; 68
utility theory; 75; 153

V

value of a statistical life; 27; 29
variance; 29; 30
variation; 39; 40; 84; 97; 143; 145
vessel; 94; 121; 130; 131; 138
vision; 110
vulnerability; vi; 79; 88; 140; 151; 156

W

well stream; 12; 48; 49
willingness to pay; 27; 79; 152

Z

zero vision; 109; 110

Printed in the United Kingdom
by Lightning Source UK Ltd.
118200UK00007B/4